U0170220

陈月巧◎著

Colorful Guizhou in Festival Costumes

图书在版编目（CIP）数据

盛装里的多彩贵州 / 陈月巧著. --北京：中国书籍出版社，2022.1
ISBN 978-7-5068-8835-6

Ⅰ.①盛… Ⅱ.①陈… Ⅲ.①少数民族—民族服饰—服饰文化—介绍—贵州 Ⅳ.①TS941.742.8

中国版本图书馆CIP数据核字（2021）第232217号

盛装里的多彩贵州

陈月巧　著

责任编辑	牛　超
责任印制	孙马飞　马　芝
封面设计	中尚图
出版发行	中国书籍出版社
地　　址	北京市丰台区三路居路 97 号（邮编：100073）
电　　话	（010）52257143（总编室）（010）52257140（发行部）
电子邮箱	eo@chinabp.com.cn
经　　销	全国新华书店
印　　刷	炫彩（天津）印刷有限责任公司
开　　本	710 毫米 × 1000 毫米　1/16
字　　数	215千字
印　　张	16
版　　次	2022 年 1 月第 1 版　2022 年 1 月第 1 次印刷
书　　号	ISBN 978-7-5068-8835-6
定　　价	69.00 元

版权所有　翻印必究

博物馆展厅

前言

贵州有48个少数民族，其中有17个世居少数民族，以苗、侗、布依等少数民族为主，基本都保留并使用着传统民族服饰，款式有近300种之多，这些种类繁多的服饰几乎涵盖了中国苗族历史上所有服饰的形制、工艺及材料。

贵州少数民族服装及其附属饰品分类较多，以下几种分类较为常见。按用途分为：一等盛装，重大节日或婚庆时穿，这类服饰制作一般要耗时数年甚至十几年，并附有大量银饰品相配；二等盛装，出席较重要的社交活动，如走亲访友、年轻人约会时穿；常装和劳动装，常装的领、袖处有少量刺绣，劳动装则为素衣，都以衣裤为主；按工艺分为：刺绣、蜡防染、织锦等；按年龄分为：老年装、中年装、青年装和儿童装等；按材料分为：麻、棉、毛、丝、皮等；按型制分为裙装、衣裤装。尽管分类方式方法众多，但最直接、最有效的识别和分类就是按地区、民族来分。本书以贵州五彩黔艺民族服饰博物馆2016年开馆以来的展陈品为载体，按照贵州九个市州代表性服饰介绍划分章节，既展现了盛装里的多彩贵州民族文化，也是对五彩黔艺近五年展览的时代记录。在展馆即将搬迁的遗憾中，以纸媒为介，唤起人们对这批藏品的珍贵回忆。在新馆展陈中，会陈列更多未展出的藏品，犹如翻书，一页一章，一展一题，把沉睡在库房里的藏品逐步唤醒。

贵州大多数少数民族没有文字，本民族的历史记录大部分依托妇女手中的针线，染织绣成各种特定的图案，留在可以轻易转移的衣物上，让这些本民族的历史，化成一种文化符号，穿在身上，世代相传，永不丢失。

盛装里的多彩贵州

Introduction

In Guizhou Province, there are 48 ethnic groups, 17 of which have lived for generations, mainly Miao, Dong, Puyi and other ethnic groups. Almost all of them have preserved and used traditional national costumes with nearly 300 kinds of styles. These various costumes cover almost all the shapes, crafts and materials of the costumes in Chinese Miao history.

There are many classifications of Guizhou national costumes and ornaments, with the following classification relatively common. According to the usage, it can be divided into three kinds: first-class festival costume, worn at important festivals or weddings, usually taking several years, or even more than ten years to make, with a lot of silver ornaments to match; second-class festival costume, worn when attending more important social activities, including visiting relatives and young people dating; daily dress, also known as casual dress or labor dress, usually with a small amount of embroidery on collar and sleeves, while labor dress, the plain clothes, mainly composed of a jacket and pants. It can be divided into embroidery, batik and brocade etc. according to handicraft art. According to age, it can be divided into old-aged costume, middle-aged costume, youth's costume and children's costume. According to the material, it can be divided into: linen, cotton, wool, silk and leather etc.. According to the shape type, it is divided into skirt and trousers. Although there are many ways of classification, the most direct and effective way to identify and classify is by region and nationality.

Based on the collection of Guizhou Multi-color Art Museum, taking the exhibits displayed since 2016, the opening of the museum, as the carrier and the representative costumes of nine prefectures and cities in Guizhou as the support, this book provides a window to appreciate

the colorful Guizhou national culture in the costumes, which is also the time record of the five-year exhibition of the museum. In the sadness of the upcoming relocation of the exhibition hall, the paper media is used to evoke people's precious memories of the collections. In the new exhibition hall, there will be more un-exhibited collections, just like turning a book, page by page, chapter by chapter. The collections will be exhibited by topic and gradually wake the dormant collections in the storeroom.

Most ethnic groups in Guizhou have no written languages. Their historical records mainly rely on the needles and threads in women's hands. They are dyed, weaved and embroidered into various specific patterns on costumes that can be easily transferred. These ethnic histories can be transformed into a cultural symbol on costumes and kept from generation to generation.

序

余未人

文学与民族民间文化，不知不觉间进入了我数十年的日子，丰富而浩瀚。精筛细选，留下了一个个有故事的鲜活生命。其中的一片靓丽，当是贵州少数民族服饰。

进得五彩黔艺博物馆，犹如踏进一片花海。游入花海细观细品，一枝一朵，一瓣一蕊，馥郁芬芳，暗香真色。

这些如云似霞的霓裳羽衣，是民间绣娘铢积寸累而成。破线绣的绣制过程让人讶异。一根普通丝线，被剖成数股极细的丝，用鲜皂荚捋过，瞬间嬗变，丝丝光洁秀雅。绣娘飞针走线，将巧夺天工的纹样一一绣出。纹样并非传统的写实，而是经代代传承铭刻于绣娘心中的苗族神话故事，那超现实的想象颇具现代派的意蕴。我见过的传统衣饰上，绣娘会将人物和动物的肠子绣出来，仿佛是肉眼的透视。因为传说肠子是一身灵气之所在，绣娘们不会忽略。它在施洞古老的苗族盛装上，有着完美体现；而在当代传承链中，这一纹样几近消失。

哪一种绣法最难？我原以为就是破线绣了。可我在施洞一带察访听年迈的绣娘说，最难做的是堆绣。先要将绸缎染成各种颜色，剪成小长条，再折成带尾的三角形，由里向外堆叠，要在每个三角形的顶尖和底部各缝钉一针，固定成型后，不断堆叠拼成各种纹样，以鱼和鸟的纹样最为常见。单是要将那些小三角做得一模一样，就颇为不凡。一件盛装的制作，常常要花费数年的功夫。这也是绣娘的骄傲和乐趣。

记得儿时养蚕，曾经尝试过让蚕宝宝吐丝在一块木板上，成为一片丝布。而板丝绣，就是用它作为底布。板丝绣是苗族先民的创造，真正的纯天然。舟溪等地的盛装中有它，而集中体现板丝绣特色的，有丹寨雅灰一代的苗族鼓藏服，它是男女皆穿的百鸟衣。在染好的成片板丝上，绣有象征祖神的鹡宇鸟，各种形态的鸟纹布满服装，不论从材质或纹样上，都独具特色。

用金属锡来当线，做成的绣品是锡绣。锡绣主要流行于剑河县清水江沿岸的部分高山苗寨。这是世上独一无二的金属挑花绣。我曾经在不通公路的剑河展留村见识过锡绣制作过程。在一块挑花的家织布上，姑娘先用剪刀把锡片剪成丝线一样的细丝，这需要极好的眼力和过硬的剪刀功夫。我观看时屏住呼吸，生怕她会受扰而剪断了锡丝。姑娘终于剪下来了，随即将闪亮的锡丝绣缀于图案中。一块锡绣的诞生，积淀着悠长的年华和精湛的艺术。

挑花技艺普遍应用于各民族服饰中，看成品似乎像十字绣。但民族民间的挑花不需画底稿，先挑出图案的外轮廓，再逐步丰富内纹，最后填充细部。绣娘还采用了“反面挑花正面看”的特殊技法。绣娘追求的不只是成品，还包含刺绣的过程。这种技艺，就超越十字绣了。

马尾绣主要流行于水族地区，在苗族、布依族那儿，我也见过。二十年前，我在三都见到几匹秃尾巴马。原来，那白色的马尾是被女人们剪去做马尾绣了。在三都，我又看到几条清代古老的马尾绣背扇，用它与现代流行的背扇相比较后发现，在清代，黄色是皇家的专用色，民间不得擅用。而水族生活的这一隅“山高皇帝远”，水族女人们全然不在乎朝廷的严规，一个个把背扇都做成了灿烂的金黄，上面再配洁白的马尾绣，极有审美眼光。新中国成立后，红色成为流行色，水族背扇又大多改用了红色，只有“点睛”的马尾绣依然是白色；红白相间的这种配色方式，使得绣品更加喜庆。

织锦是服饰的一大元素。苗族织锦的配色艳丽而讲究，织锦方法分为挑织和编织，宽的有尺许，窄的只有一寸。头帕、背带、裙带、腰带等，用上织锦就显得典雅飘逸，风韵卓然。在纹样上，织

机用丝的经纬走向决定了织锦的图案。因为织机经纬的限制，它造就了一种特殊的美——这些原本就富有想象力的图案直接被几何图形化、抽象化了。苗锦将苗族古歌中的花鸟虫鱼、飞禽走兽都做成几何纹样，这与现代艺术特别契合，别有一番韵味。其实现代艺术中，就有许多原始艺术的基因。织锦纹样的艺术夸张，可以说是亘古传承下来的、直通现代审美的“原始美”。

蜡染是民族民间服饰的半壁江山。蜡染与蓝印花布不同，它没有版，全是手绘。所以，你见不到两件一模一样的作品。这种纯手工，件件独一无二。苗族蜡染的图案生动明丽，独具特色。纹样既是写实的，又是抽象的；在布局上既是对称的，又有许多突破，特别灵动。苗族的民间信仰是万物有灵，飞天入地的丰富想象力，在蜡画中处处有所体现。民间传统做蜡染，追求完整、工整、无瑕疵，这是祖祖辈辈染娘追求的最高境界；而艺术家和现代人却从“瑕疵”中发现了另一种美，那是由蜡天然开裂而形成的冰纹，这种冰纹古朴典雅自然，有抽象画的意味，是蜡染美的特色之一。

上述种种，仅为有代表性的一些服饰元素，在五彩黔艺博物馆的藏品中处处可见，在其文字说明中有详叙，我就不赘述了。

进得此馆，是一种美的享受。“此身不知栖何处，天下人间一片云。”而当你细观深品，就会发现，博物馆需要文化的浸润、学术的熏陶和虔诚细致的经营。尤其是民族服饰，大多是在民间几经曲折，才辗转来此处安身立命，展示于人。藏品中蕴含的不凡经历和故事，其隐藏的学术价值，是博物馆一笔巨大的无形财富。

如今，“五彩黔艺”藏品日渐丰富，学术价值与日俱长。然而前路漫漫，大物小事永无止境。展望明天，五彩黔艺定能进一步弘扬这批无形财富的精气神。

Preface

Yu Weiren

Literature and folk culture, imperceptibly into my life for decades of years, are rich and vast. Through fine selection, stories full of fresh lives are kept. One of the beautiful things is Guizhou ethnic group's costumes and ornaments.

Entering Multi-color Art Museum is like stepping into a sea of flowers. Every detail of the costumes and ornaments, like one branch, one petal or one pistil, has its elegance and fragrance.

These colorful and feathered costumes are made by common embroidering women through their inheritance and practice. The embroidery process of thread-splitting stitch is amazing by splitting a common thread into several fine threads and rubbing them with juice of fresh Chinese honey locust to make them smooth. The embroidering women stitch the ingenious patterns one by one with a small needle in hand. The patterns are not traditional realistic objects, but Miao's fairy tales engraved in minds of embroidering women through generations. The surreal imagination has a modernist connotation. I have once seen that embroidering women stitched the intestines of humans and animals on traditional clothes, as if it were the perspective of the naked eye. Because it is said that intestines are the place of a body's aura, embroidering women will not ignore it. Such patterns were fully presented on traditional Miao's festival costumes at Shidong Town, but almost disappeared from contemporary inheritance.

Which one is the most difficult? I thought it was thread-splitting

embroidery. But according to elder embroidery women when I did field research at Shidong area, the most difficult thing is piling embroidery. At first, the brocade is dyed into different colors and cut into straps, which are folded into triangles with tails. which are piled from inside to outside with a stitch respectively at the corner and at the bottom. After that, different patterns, with fish and birds as common ones, are piled and sewed. It's not easy to make all the small triangles the same. Therefore, it takes several years to make a festival costume, which is also the pride and fun for embroidering women.

I remember that when I was a child, I tried to make my silkworms spin on a board to form a silk cloth, which is just the base cloth for board embroidery however. It's the creation of Miao people's ancestors and really natural. It is presented in the festival costumes at Zhouxi areas. The Yahui style, "hundred-bird costume" in Danzhai, embodies the typical characteristics of board embroidery. "Jiyu" birds, symbolizing the ancestor god, are embroidered on the dyed board silk and bird patterns of various shapes are all over the costume. They are unique no matter from the material or the pattern.

The embroidery work made of metal tin as thread is called tin embroidery, which is popular in some Miao villages in mountains along Qingshui River of Jianhe County. It is the unique metal cross-stitch. I have seen tin embroidering process at Zhanliu Village of Jianhe County, which is less accessible by road. The embroidery girl first cut the tin sheet into silk-like filaments with scissors, which is out of great and exemplary skill. Then she embroidered patterns on a cross-stitched home-woven cloth with shining tin threads. The birth of a piece of tin embroidery work has accumulated a long time and exquisite art.

Cross-stitch technique is commonly used in national costumes and ornaments. The cross-stitched work looks like "cross embroidery". However, it is not necessary to draw a draft for cross-stitch. At first, the outlines of patterns are cross stitched. Then the inner patterns are gradually enriched with the details filled as the last step. Embroidery women apply a special technique by crossing stitch at the back and looking from the front. Embroidery women pursue not only finished products, but also the process of embroidery. This technique goes beyond cross stitch.

Ponytail embroidery is mainly popular in Shui residential areas. I have seen it in Miao and Puyi residential areas too. Twenty years ago, I saw several bald tailed horses in Sandu County. It turned out the white horsetail was cut by women to do ponytail embroidery. In Sandu county, I have seen several ancient baby carriers in ponytail stitch from Qing Dynasty. Compared with popular baby carriers in contemporary times, I found that Shui women use shining yellow or golden color used by the emperors only in Qing Dynasty to make baby carriers stitched with white ponytails. After the founding of the People's Republic of China, red has become a popular color, and most of the baby carriers of Shui people have been changed to red, only keeping the "finishing point" white. The color matching of red and white makes the embroidery more festive.

Brocade is a major element of costumes. With gorgeous and exquisite colors and different width size, brocade can be divided into picking weaving and weaving according to its methods. With brocade, headscarf, baby carrier, sash of skirt and sash appear elegant and charming. The thread longitudinal and latitudinal directions of loom determine the patterns of brocade and create a special beauty——these original imaginative patterns are directly geometrically patterned and abstracted. The flowers, insects, fish, birds and animals in Miao's ancient songs are brocaded into geometric patterns, which particularly agrees with modern art and has a unique charm. In fact, there are many sources of primitive art in modern art. The artistic exaggeration of brocaded patterns can be said to be the "primitive beauty" inherited from ancient times and directly connected with modern aesthetics.

Batik is half of the national costumes. Batik is different from blue calico. It has no printing plate and is all hand-painted, so you can't see two identical batik works. Each piece of this kind of pure handwork is unique. The patterns of Miao batik are vivid and distinctive. Patterns are both realistic and abstract. The flexible layout is not only symmetrical, but also has many breakthroughs. Miao people's folk belief is that everything has a spirit. The rich imagination is reflected in batik patterns everywhere.

The traditional batik technique pursues integrity, neatness and flawlessness, which are the supreme goal pursued by generations of batik craftswomen. However, artists and modern people have found another beauty from the "flaw", which is the ice patterns formed by the natural cracking of wax. This kind of ice pattern is simple, elegant and natural, with the meaning of abstract painting, which is one of the characteristics of batik beauty.

All of the above are only representative elements of costumes, which can be seen everywhere in the collection of Multi-color Art Museum. I will not repeat them here since they are demonstrated in detail in the text description of each collection.

Entering this museum is to enjoy the beauty of national costumes. When you take a closer look at the costumes, you will find that this museum is culturally infiltrated, academically nurtured and piously and meticulously managed. In particular, most of the national costumes exhibited in the museum have their own extraordinary experiences and stories. The invisible academic value of the collection is a huge intangible wealth of the museum.

Nowadays, the collection of Multi-color Art Museum is increasingly rich, and the academic quality is growing with each passing day. However, there is a long way to go. Looking forward to tomorrow, Multi-color Art Museum will be able to further promote the spirit of the intangible wealth.

目录

Contents

第一章
贵阳地区少数民族服饰

Chapter 1 Ethnic Group's Costumes and Ornaments in Guiyang Area

» 名称：贵阳乌当下坝苗族旗子领式贯首女套装

» 时代：约1980年

» 馆藏：W.1732

主要流行于贵阳、清镇、龙里、修文等地，以贵阳乌当下坝为代表。该套装由上衣、百褶裙、围腰、头帕组成。妇女发式为绾髻于顶，用约十八米长的青布头帕呈人字形交叉缠成盘状。盛装时上衣为青布旗子领式贯首衣，后背、双袖挑花，布贴方块纹、卷草纹等图案，图案间钉梅花银泡，并以醒目的蓝色镶边突出双袖、胸襟花块。下装为青布贴布绣缀花拼接成百褶裙，另有蜡染和彩布混合的围腰。苗族贯首服是人类早期服饰款式的遗风，穿着时从头往下套，因而得名，分为无领式、翻领式和旗子领式三种，此款为旗子领式。相传该支系苗族女子双袖、胸襟上的方块图案象征祖先蚩尤九黎的符号。苗族是一个迁徙的民族，为了避免战乱，易于识别，以苗族祖先蚩尤九黎作为民族标志，既表达对祖先的缅怀之情，又起到加强民族凝聚力的作用。

图1-1-1　贵阳乌当下坝苗族旗子领式贯首女套装

Picture 1-1-1　Miao's Through-hole Woman Outfit with Tie-like Straps Welted on Front Border at Xiaba in Wudang District of Guiyang City

» Name: Miao's Through-hole Woman Outfit with Tie-like Straps Welted on Front Border at Xiaba in Wudang District of Guiyang City

» Time: About 1980

» Collection: W.1732

Represented by the style at Xiaba Village of Wudang District of Guiyang City, the outfit is distributed in Guiyang City, Qingzhen City, Longli County and Xiuwen County in Guizhou Province. It is composed of jacket, pleated skirt, apron and headscarf. Women use a 18m-long black cloth to wrap their hair in the shape of a reversed V. When dressed up, women wear black pullover with cross-stitched patterns on back and sleeves, cloth-stitched patterns of square and curly grass, and decorated with silver bubble in the shape of plum flower and blue bindings to highlight the sleeves and front. With colorful batik apron, women wear black-cloth made pleated skirt with patterns of flowers. Originating from early clothes style, Miao nationality's pullover is named for its wearing feature: dressed through the head. The outfit has three styles: collarless style, style with lapels and style with tie-like straps welted on front border, with the last

图1-1-2　套装正面展示

Picture 1-1-2　Front Show

图1-1-3　套装背面展示

Picture 1-1-3　Back Show

one represented in the picture.

In legends, square patterns on sleeves and front pieces of woman's outfits symbolize the sign of the mythological warrior "Chiyou", the ancestor of Miao. The sign is regarded as a symbol for Miao's identification in migration during the war, which not only commemorates Miao's ancestors, but also strengthens solidarity.

图1-1-4　身着盛装的男子和女子跳苗族花棍舞（陈月巧　摄）

Picture 1-1-4　Dressed Men and Women Dancing the Miao "Huagun Dance" (Dance with Ornamented Sticks) (Photographed by Chen Yueqiao)

» 名称：贵阳花溪苗族无领挑花贯首女套装
» 时代：约1990年
» 馆藏：W.1731

主要流行于贵阳市花溪区、平坝马场等地，以贵阳花溪为代表。该套装由上衣、百褶裙、前围腰、后围腰、背牌、头帕组成。上衣为家织平纹青布无领贯首服，双袖、前胸、后背及后衣下摆挑绣几何纹屋架花和麦穗花。其形制是将一整块布折成衣胸、衣背两大片，然后在对折处剪开为领口，再将两块布边连缀成了衣身，下装为青布百褶裙，裙前系挑花几何纹和花纹前围腰，裙后系挑花飘带围腰，头帕为织锦宽花带。另有方形挑花背牌，背牌花带交叉于胸前。苗族贯首服是人类早期服饰款式的遗风，分为无领型、翻领型和旗子领型三种，贵阳花溪苗族女套装多为翻领型，此款为无领型。盛装时，头插七八根银簪子，并戴银头花。

图1-2-1　贵阳花溪苗族无领挑花贯首女套装（正面）

Picture 1-2-1　Miao's Collarless Woman Pullover in Huaxi District of Guiyang City (Front)

图1-2-2　贵阳花溪苗族无领挑花贯首女套装（背面）

Picture 1-2-2　Miao's Collarless Woman Pullover in Huaxi District of Guiyang City (Back)

» Name: Miao's Collarless Woman Pullover in Huaxi District of Guiyang City

» Time: About 1990

» Collection: W.1731

Represented by the style in Huaxi District of Guiyang City, the outfit is popular in Huaxi of Guiyang City and Machang of Pingba County. It is composed of jacket, pleated skirt, front apron, back apron, long shoulder aprons with tassels and headscarf. The black-cloth made pullover is decorated with cross-stitched geometrical patterns on sleeves, front, back and lower edges. The jacket is tailored with a cloth folded into two pieces and a hole cut at the folding point as the collar hole. Wearing brocade headscarves, cross-stitched front aprons with geometrics and flower patterns and cross-stitched streamers hanging at the back aprons, women are dressed in black-cloth made pleated skirts and square cross-stitched back ornamentation with tassels crossed in front.

图1-2-3　挑花前围腰

Picture 1-2-3　Cross-stitched Front Apron

Originating from early clothes style, Miao's pullover has three styles: collarless style, style with lapels and style with flag-like collars. Style with lapels is common in Huaxi District and the picture shows the collarless style.

Women decorate their topknots with several silver hairpins and silver flowers in festival costumes.

图1-2-4 挑花飘带后围腰

Picture 1-2-4 Cross-stitched Back Apron with Streamers

图1-2-5　挑花几何纹

Picture 1-2-5　Cross-stitched Geometrics

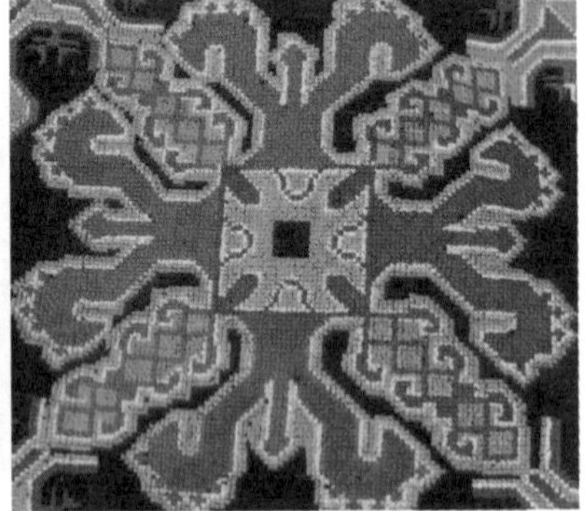

图1-2-6　挑花几何纹屋架花图案

Picture 1-2-6　Geometrical Roof Frame Pattern in Cross-stitch

图1-2-7　挑花几何纹麦穗花图案

Picture 1-2-7　Geometrical Wheat Pattern in Cross-stitch

图1-2-8　身着盛装的女子载歌载舞

Picture 1-2-8　Women Dressed in Festival Costumes Singing and Dancing

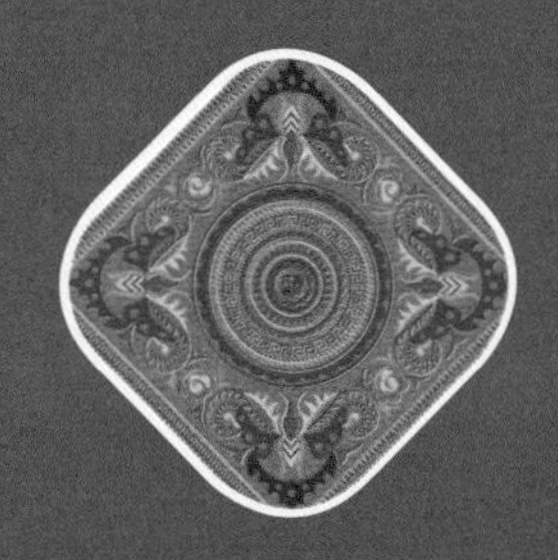

第二章
黔东南地区少数民族服饰

Chapter 2 Ethnic Group's Costumes and Ornaments in South-East of Guizhou Province

» 名称：贵州台江苗族破线绣对襟女套装
» 时代：约1980年
» 馆藏：W.1749

主要流行于贵州台江、镇远、施秉等地，以台江施洞为代表。该套装由上衣、百褶裙、围腰、头帕等组成。上衣用亮布制作，双袖、肩部、前襟主要是刺绣和织锦工艺，施洞地区盛装以红色和蓝紫色为主，绣法以破线绣和数纱绣居多。图案以当地流传的苗族古歌中传唱的历史故事和神话传说为主，还有各种变形的鸟、蝴蝶、龙、象等，其中破线绣绣工精湛，被誉为苗绣中最精美的刺绣之一，下装为用亮布制作的百褶裙。不穿戴银饰时，配织锦围腰和织锦头帕。该套装多为未婚女子的盛装，配双龙抢宝银头饰、银头花、绞丝银项圈、龙项圈、银手镯，衣服饰银衣片，脚穿绣花鞋，冬天时还有袖笼。

图2-1-1 贵州台江苗族破线绣对襟女套装（正面）

Picture 2-1-1 Miao's Front Symmetric Woman Outfit in Thread-splitting Stitch in Taijiang of Guizhou (Front)

» Name: Miao's Front Symmetric Woman Outfit in Thread-splitting Stitch in Taijiang of Guizhou
» Time: About 1980
» Collection: W.1749

Represented by the style in Shidong Town of Taijiang County in Guizhou Province, the outfit is popular in Taijiang, Zhenyuan and Shibing. It is composed of jacket, pleated skirt, apron and headscarf. The jacket is made of bright cloth with embroidery and brocade on sleeves, shoulders and front lapels. The festival costume is red and blue-purple with patterns of transformed bird, butterfly, dragon and elephant in historical and fairy tales in yarn-counting stitch and thread-splitting stitch which is praised as the most exquisite embroidery. Wearing brocaded apron and headscarf, women are dressed in pleated skirts made of bright cloth. This festival costume is adorned with many silver ornaments like headdress of "two dragons chasing after a pearl", headdress flowers, neck rings, dragon neck rings, bracelets, sequins on jacket, embroidered shoes and loose sleeves in winter.

图2-1-2　贵州台江苗族破线绣对襟女套装（背面）
Picture 2-1-2　Miao's Front Symmetric Woman Outfit in Thread-splitting Stitch in Taijiang of Guizhou (Back)

图2-1-3 破线绣绣花鞋

Picture 2-1-3 Embroidered Shoes in Thread-splitting Stitch

图2-1-4 亮布对襟上衣背面

Picture 2-1-4 Front Symmetric Outfit Made of Bright Cloth（Back）

图2-1-5　施洞苗族女盛装正反面（陈月巧　摄）

Picture 2-1-5　Miao Women's Festival Costume at Shidong Town （Front and Back）(Photographed by Chen Yueqiao)

» 名称：雷山西江苗族绉绣对襟女套装
» 时代：约1980年
» 馆藏：W.2160

主要流行于贵州雷山、台江、凯里等地，以雷山西江为代表。该套装由上衣、百褶裙、飘带裙等组成。上衣为家织青布对襟衣，双袖、门襟、肩部红底蓝、绿绉绣关于久保的故事及花草纹，并饰亮片。下装为青布百褶长裙，长裙外还有刺绣飘带裙，共27根飘带，飘带一般为单数，可根据人体胖瘦决定飘带数量，每根飘带又由五个部分连接而成，使用双针绣、打籽绣等工艺绣的各种动物纹、花纹，所有图案不重复，甚是精致。双袖上用辫绣工艺绣有龙纹和人纹，图案是苗族古歌里清水江一带龙船节的来历，即久保的故事。传说很久以前，清水江里有条恶龙，它吃掉了保公的儿子久保，保公愤而杀死恶龙，为民除了害。恶龙灵魂幡然悔悟，托梦给人们依龙形制舟，每年五月举行竞渡，以保风调雨顺。从此清水江边的苗族都会过每年一度的传统龙舟节。女子盛装参加龙舟节，头戴大银角、银头花、银帽、银梳、银项圈、银锁牌、银手镯，衣服上饰银衣片。该上衣制作时间大约在1970年，飘带裙制作时间大约在2000年。

图2-2-1 雷山西江苗族绉绣对襟女套装

Picture 2-2-1 Miao's Front Symmetric Woman Outfit in Wrinkling Stitch at Xijiang, Leishan

» Name: Miao's Front Symmetric Woman Outfit in Wrinkling Stitch at Xijiang, Leishan

» Time: About 1980

» Collection: W.2160

Represented by the style in Xijiang Town of Leishan County in Guizhou Province, the outfit is popular in Leishan,Taijiang, and Kaili. It is composed of jacket, pleated skirt, skirt with streamers, etc.. Made of home-woven black cloth and decorated with silver ornaments, the jacket is buttoned on the right embroidered with tales of Jiubao and patterns of flowers and grasses on blue and green sleeves and front lapels in wrinkling stitch with red. The black pleated long skirt usually has streamers in odd number, with exquisite patterns of animals and flowers in double needle stitch and grain stitch on each streamer which is combined with five parts. The dragon and human pattern in braid stitch on the sleeves is a tale about the origin of "Dragon Boat Festival" in Qingshui River area, also a story of Jiubao. According to the legend, an evil dragon who lived in Qingshui River ate the son of Baogong, so that Baogong became so angry to kill the dragon for the people. Afterward, the dragon regretted for his actions and appeared in the people's dream in which it inspired people to build boats which look like a dragon and have competition in May every year for ensuring a favorable weather whole year. Women take part in Dragon Boat Festival in festival costumes, decorated with silver ornaments: crowns, headdresses, hats, combs, necklaces, lock, bracelets and pieces. The jacket was made in 1970 or so and the streamer skirt in 2000 or so.

图2-2-2　家织青布对襟衣上衣

Picture 2-2-2　Front Symmetric Outfit Made of Home-woven Black-cloth

图2-2-3　雷山西江苗族绉绣对襟女盛装（陈月巧　摄）

Picture 2-2-3　Miao's Front Symmetric Woman Outfit in Wrinkling Stitch at Xijiang, Leishan（Photographed by Chen Yueqiao）

» 名称：凯里舟溪苗族板丝绣对襟女套装
» 时代：约1990年
» 馆藏：W.1738

主要流行于贵州凯里、丹寨、麻江等地，以凯里舟溪为代表。该套装由上衣、百褶裙、围腰、头套、袖笼等组成。上衣为褐色缎面大袖对襟衣，双袖红、绿相间板丝绣象征宇宙天体的神秘图案。下装为青布两片式百褶短裙，围腰与上衣花式相同，下摆饰蚕丝绸彩条流苏。另有插针绣月季花黑绒布作底的头套和蓝色纯棉底的袖笼，饰银泡的背饰带。两肩和围腰，多用红色丝绸为底，以染成黄、蓝、绿等平面茧为原料剪成各种图案，再以马尾绣为框钉缝成边，形成太阳纹、牛角纹、铜鼓纹、十字纹等。盛装时，女子戴银抹额、银头花、银梳、银项圈，衣上饰银衣片、银腰带等。

图2-3-1　凯里舟溪苗族板丝绣对襟女套装

Picture 2-3-1　Miao's Symmetric Woman outfit in Board Silk Stitch at Zhouxi of Kaili

图2-3-2　插针绣月季花黑绒布头套

Picture 2-3-2　Pin Embroidered Black Velvet Headgear with Rosa Chinensis

图2-3-3　插针绣月季花蓝棉布底袖笼

Picture 2-3-3　Pin Embroidered Blue Cotton Loose Sleeves with Rosa Chinensis

» Name: Miao's Symmetric Woman outfit in Board Silk Stitch at Zhouxi of Kaili

» Time: About 1990

» Collection: W.1738

Represented by the style in Zhouxi Town of Kiali City, the outfit is popular in Kaili City, Danzhai County and Majiang County in Guizhou Province. The outfit is composed of jacket, pleated skirt, apron, headgear and loose sleeves. The jacket is a brown silk symmetric garment with board silk stitched mysterious patterns symbolizing universe and heavenly body on loose sleeves. The pleated skirt is made up of two pieces of black cloth, covered by an apron with the same patterns on jacket and ornamented with colorful tassels as lower hem. Women usually wear pin embroidered black velvet headgears, blue-cotton loose sleeves with rosa chinensis and back sashes ornamented with silver bubbles. The yellow, blue and green plate cocoons are cut into different patterns of sun, horn, bronze drum and cross on red silk, with ponytail stitch frame as edge. Women also wear silver ornaments including headband, headdress, comb, neck rings, pieces and sash at festivals.

图2-3-4 象征宇宙天体的神秘图案

Picture 2-3-4 Mysterious Patterns Symbolizing Universe and Heavenly Body

图2-3-5 蚕丝绸彩色流苏

Picture 2-3-5 Colorful Silk Tassels

图2-3-6　亮布百褶短裙

Picture 2-3-6　Pleated Skirt Made of Bright Cloth

图2-3-7　丽人出山寨（王绍帅　摄）

Picture 2-3-7　Young Ladies Walking out of the Cottage (Photographed by Wang Shaoshuai)

» 名称：台江革一苗族三角堆绣交襟女套装

» 时代：约1980年

» 馆藏：W.1763

主要流行于贵州台江、凯里等地，以台江革一为代表。该套装由上衣、百褶裙、绸缎头帕等组成。妇女穿着常服时，头包藏青土布头帕，帕上常绣花饰，上衣为红色绸缎交襟服，衣领、前襟、双袖布堆绣鱼纹、鸟纹、方块纹，后背上方布堆绣八鱼合欢图，下方三角堆绣两只栩栩如生的锦鸡，衣左右及下摆锦缎饰之。下装为红、黑、青、蓝绸缎相间的百褶裙，裙下摆饰锦缎。台江革一的堆绣是苗族刺绣中极其复杂且耗工耗时的技艺之一。制作前要先将绸缎染

图2-4-1　台江革一苗族三角堆绣交襟女套装

Picture 2-4-1　Miao's Woman Outfit with Cross-button in Piling Stitch at Geyi of Taijiang

成各种颜色，剪成小长条，再折成带尾三角形，大图案的可以先剪纸，将剪纸贴在绣布上，根据纸花图案由里向外堆叠，每堆一个三角形就要在三角形的顶尖缝钉一针，在三角形底部堆叠处再缝钉一针，固定成型后不断堆叠拼成各种几何纹或动物图案，然后再将绣片缝钉在衣服特定的位置上。盛装时，女子戴银项圈、银胸饰、银花帽、大银角，衣服饰银衣片。

图2-4-2　红色绸缎上衣

Picture 2-4-2　Red Brocaded Jacket

» Name: Miao's Woman Outfit with Cross-button in Piling Stitch at Geyi of Taijiang
» Time: About 1980
» Collection: W. 1763

Represented by the style at Geyi Town of Taijiang County, the outfit is popular in Taijiang County and Kaili City in Guizhou Province. The outfit is composed of jacket, pleated skirt and brocaded headscarf. In daily dress, women wear towel turbans with embroidered flower patterns and red brocade jackets with cross-button. The collars, front lapels, and sleeves are piling stitched with patters of fish, bird, square prism, "eight fish in harmony around a Hehuan flower"on the upper part, two vivid cocks on the bottom part of the back and brocaded hems, which match with pleated skirts alternated in red, black, green and blue with brocaded hems. Women wear many kinds of silver ornaments including neck rings, corsage, turban, crown and sequins at festivals.

图2-4-3 三角堆绣八鱼合欢图案
Picture 2-4-3 Pattern of "Eight Fish in Harmony Around a Hehuan Flower" in Piling Stitch

图2-4-4　三角堆绣公鸡图案

Picture 2-4-4　Pattern of Cock in Piling Stitch

图2-4-5　堆绣方形图案

Picture 2-4-5　Pattern of Square Prism in Piling Stitch

图2-4-6　堆绣、打籽绣衣领局部图

Picture 2-4-6　Collar Panel in Piling and Grain Stitch （Section）

图2-4-7　红、黑、青、蓝色绸缎相间百褶裙局部图

Picture 2-4-7　Pleated Brocaded Skirt Panel Alternated in Red, Black, Green and Blue （Section）

图2-4-8 一路欢笑一路歌（王济文 摄）

Picture 2-4-8 Laughing and Singing All the Way (Photographed by Wang Jiwen)

» 名称：黄平重安江僅家蜡染对襟女套装

» 时代：约2000年

» 馆藏：W.1735

主要流行于贵州黄平县境内，以黄平重安江沿岸的村寨为代表。该套装由上衣、百褶裙、围腰、头帕、绑腿等组成。上衣为蜡染对襟衣，双袖、后衣下摆彩线挑绣几何纹，左右门襟及后背蜡染铜鼓纹和卷云纹。下装为青布百褶短裙，裙子的二分之一处缀蜡染镶红色织锦彩带褶皱裙边。另应配有蜡染围腰、蜡染红缨帽、绑腿和织锦带。妇女绾髻于顶，外戴红缨帽。僅家自称是上古传说中射日英雄后羿的后代。黄平重安江 僅家蜡染因其线条流畅、表现手法细腻，在贵州蜡染中颇具代表性，图案以变形的蝴蝶纹、鸟纹、缠枝纹和铜鼓纹居多，表现了 僅家人对自然的崇拜和对繁衍子孙的殷切希望。

图2-5-1　黄平重安江僅家蜡染对襟女套装

Picture 2-5-1　Gejia's Front Symmetric Woman Batik Outfit along Chong'an River, Huangping of Guizhou

» Name: Gejia's Front Symmetric Woman Batik Outfit along Chong'an River, Huangping of Guizhou

» Time: About 2000

» Collection: W.1735

Represented by the style at villages along Chong'an River, the outfit is popular in Huangping County in Guizhou Province. The outfit is composed of jacket, pleated skirt, apron, headscarf and leggings. The top is a front symmetric batik jacket with cross-stitched geometric patterns on the hems of sleeves and back jacket, and batik patterns of bronze drum and cirrus cloud on left and right lapels and back. The bottom is a black-cloth pleated mini skirt decorated with batik red brocaded colorful pleated hem at half. In addition, women wear batik apron, batik red-tasseled cap, leggings and brocaded sash. Women keep topknots with red-tasseled cap. People from Gejia claim to be the offspring of Houyi, the hero shooting off nine suns in Chinese tale. The batik of Gejia is representative in Guizhou batik for its smooth line and exquisite expression with patterns of transformed butterfly, bird, twines and bronze drum, which display the worship towards nature and hope for reproduction.

图2-5-2　上衣局部图

Picture 2-5-2　Jacket（Section）

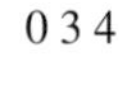

图2-5-3 衣袖局部图

Picture 2-5-3 Sleeve（Section）

图2-5-4 蜡染卷云纹

Picture 2-5-4 Batik Patterns of Cirrus Cloud

图2-5-5 蜡染铜鼓纹

Picture 2-5-5 Batik Patterns of Bronze Drum

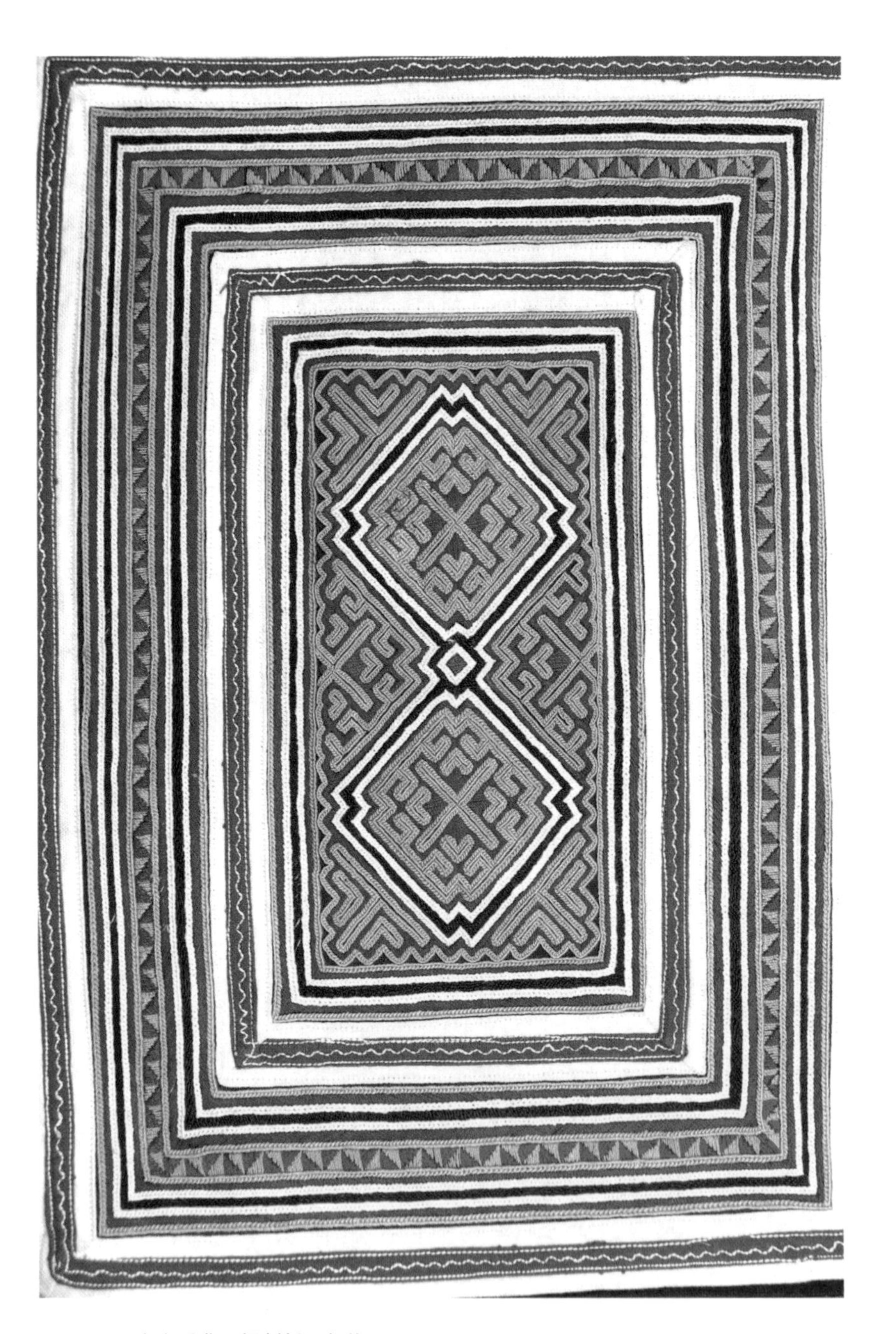

图2-5-6　上衣后背下摆刺绣几何纹

Picture 2-5-6　Cross-stitched Geometric Patterns on the Hems of Back Jacket

图2-5-7 蜡染镶红色织锦彩带褶皱裙边

Picture 2-5-7 Batik Red Brocaded Colorful Pleated Hem

图2-5-8 踩芦笙（杨秀银 摄）

Picture 2-5-8 Lusheng Dance (Photographed by Yang Xiuyin)

» 名称：凯里石龙寨西家大襟女套装
» 时代：约2000年
» 馆藏：W.1734

主要流行于贵州凯里、黄平等地，以凯里石龙寨为代表。该套装由上衣、百褶裙、围腰、头帕、绑腿等组成。上衣为蓝布大襟衣，衣襟处饰栏杆花边。下装为蜡染百褶长裙，裙身正中醒目地饰有红、黄、蓝、白几何纹方块图案。青布围腰四周饰栏杆、花边和蓝布边，上端红底剪贴绣花鸟纹。另有青布底刺绣梅花、卷云纹头帕和青布绑腿。盛装时，女子饰少量银饰。

图2-6-1　凯里石龙寨西家大襟女套装（正面）

Picture 2-6-1　Xijia's（Tentatively Designated as Miao）Woman Outfit Buttoned on the Right at Shilong Village, Kaili (Front)

» Name: Xijia's (Tentatively Designated as Miao) Woman Outfit Buttoned on the Right at Shilong Village, Kaili

» Time: About 2000

» Collection: W.1734

Represented by the style at Shilong Village in Kaili City, the outfit is popular in Kaili City and Hungping County in Guizhou Province. The outfit is composed of jacket, pleated skirt, apron, headscarf and leggings. The top is a blue-cloth batik jacket buttoned on the right with bar pattern hems on the lapels. The bottom is a batik pleated skirt decorated with red, yellow, blue and white geometric patterns in the middle. The black-cloth apron is decorated with bar patterns, hems, embroidered patterns of flowers and birds on red base. In addition, women wear headscarf with embroidered patterns of plum flowers on black-cloth base and cirrus cloud, and black-cloth leggings. Women wear some silver ornaments at festivals.

图2-6-2 凯里石龙寨西家大襟女套装（背面）

Picture 2-6-2 Xijia's（Tentatively Designated as Miao）Woman Outfit Buttoned on the Right at Shilong Village, Kaili Kaili (Back)

图2-6-3　西家故事（陈德祥　摄）

Picture 2-6-2　Xijia Story (Photographed by Chen Dexiang)

» 名称：剑河巫包苗族挑花对襟女套装
» 时代：约1990年
» 馆藏：W.1739

主要流行于贵州剑河县境内的巫包、温泉、稿旁等地，以剑河巫包乡为代表。该套装由上衣、百褶裙、围腰、绑腿等组成。上衣为对襟衣，衣襟、双袖、肩背、衣下摆处以红色为主色调挑花几何纹、鸟纹花块。下装为黑、红双色布拼接的百褶裙，红裙部分饰黄、绿、蓝三色彩条，耀眼夺目。裙上系红色丝线挑花的几何纹大围腰，下缀彩色穗子或铜钱。由于当地妇女喜用红色丝线刺绣，该服饰上的刺绣又被称为“红绣”。盛装时，女子绾髻于顶，包头帕，戴银头花，头花若三只展翅欲飞的小鸟并排立于高枝之上，另戴银锁项圈、银链、银抹额、银耳坠、银手镯，衣服上钉少量的银片。

图2-7-1 剑河巫包苗族挑花对襟女套装（正面）
Picture 2-7-1 Miao's Symmetric Woman Outfit in Cross Stitch at Wubao, Jianhe (Front)

» Name: Miao's Symmetric Woman Outfit in Cross Stitch at Wubao, Jianhe

» Time: About 1990

» Collection: W.1739

Represented by the style at Wubao Town in Jianhe County, the outfit is popular in towns of Wubao, Wenquan and Gaopang in Guizhou Province. The outfit is composed of jacket, pleated skirt, apron and leggings. The symmetric jacket is decorated with patterns of geometry and birds cross-stitched mainly in red on the lapels, sleeves, back of shoulders and hem. The batik pleated skirt is attached by red and black patches with yellow, green and blue streamers and tied by a big apron on which there are cross-stitched geometric patterns in red silk. The skirt is decorated with colorful tassels and copper coins. Such embroidery is also called "Red Embroidery" since the local women like to use red silk. Women wear headscarves and have on topknots tied with a silver hairpin as three birds on the wing lining on a branch. In addition, women wear some silver ornaments including neck rings, necklaces, headband, earrings, bracelets and sequins at festivals.

图2-7-2 剑河巫包苗族挑花对襟女套装（背面）

Picture 2-7-2 Miao's Symmetric Woman Outfit in Cross Stitch at Wubao, Jianhe (Back)

图2-7-3 上衣局部图

Picture 2-7-3 Jacket (Section)

图2-7-4 挑花几何纹、鸟纹花块

Picture 2-7-4 Cross-stitched Geometric Patterns and Bird Patterns

图2-7-5　围腰上吊白色穗子和钱吊坠

Picture 2-7-5　Apron Decorated with White Tassels and Coins

图2-7-6　挑花几何纹

Picture 2-7-6　Cross-stitched Geometric Patterns

图2-7-7　红、白、蓝、绿、黑相间的百褶裙

Picture 2-7-7　Pleated Skirt with Alternate Colors of Red, White, Blue, Green and Black

图2-7-8　剑河巫包苗族女盛装（王济文　摄）

Picture 2-7-8　Miao's Festival Costume at Wubao, Jianhe (Photographed by Wang Jiwen)

» 名称：剑河革东苗族对襟女套装

» 时代：约1970年

» 馆藏：W.1733

主要流行于贵州台江、剑河等地，以剑河革东、交东周边为代表。该套装由上衣、百褶裙、围腰、绑腿等组成。上衣为家织自染亮布对襟衣，双袖、肩、衣襟绣几何纹、菱形纹花块，另有织锦麦穗花带饰之。下装为家织青布百褶裙。另有织锦围腰和彩带绑腿。苗族古歌里有骏马飞渡故事，相传苗族迁徙途中被朝廷官兵追赶，苗族某个支系骑马飞渡江河，摆脱了围剿。苗族妇女以刺绣的方式记录了这个传说故事。盛装时，妇女绾大髻于头顶，包丝线头帕，头顶露发髻，戴多种银饰，包括银抹额、银头花，以头顶挂银质粗链形项饰为独具的特征，另戴绞丝银项圈、银锁牌、银手镯等。

图2-8-1　剑河革东苗族对襟女套装

Picture 2-8-1　Miao’s Symmetric Woman Outfit at Gedong, Jianhe

» Name: Miao's Symmetric Woman Outfit at Gedong, Jianhe

» Time: About 1970

» Collection: W.1733

Represented by the style at areas of Gedong and Jiaodong in Jianhe County, the outfit is popular in counties of Taijiang and Jianhe in Guizhou Province. The outfit is composed of jacket, pleated skirt, apron and leggings. The symmetric jacket is decorated with geometric and diamond patterns and brocade hem of ear of wheat on sleeves, shoulders and lapels. The batik pleated skirt is cross-stitched with patterns of human figures in the middle and decorated with brocade apron and colorful leggings. According to a story named "steed galloping" in Miao's song, Miao people were chased by soldiers of imperial court on the road of migration, a certain branch of Miao rode on the horses and flew across the river to get rid of encircle and suppress. Miao women embroider these beautiful stories on their cloth and record stories with beautiful patterns.

Women have on topknots tied with a silver hairpin and wear velvet headscarves. In addition, women wear some silver ornaments including headband, headdress, neck rings, locks and bracelets at festivals.

图2-8-2　衣襟局部图

Picture 2-8-2　Jacket Lapels (Section)

图2-8-3　菱形几何纹

Picture 2-8-3　Diamond Patterns

图2-8-4　织锦麦穗花带

Picture 2-8-4　Brocade Hem of Ear of Wheat

图2-8-5　百褶裙中间的人纹

Picture 2-8-5　Pattern of Human Figure in the Middle of Pleated Skirt

» 名称：剑河南寨苗族锡绣对襟女套装
» 时代：约1980年
» 馆藏：W.1741

主要流行于贵州剑河南寨、南加、敏洞、观么、九旁等乡镇，以南寨、南加为代表。该套装由上衣、百褶裙、前后围腰、头帕、绑腿等组成。上衣为家织青布对襟衣，后衣背饰有几何纹锡绣背褡。下装为青布百褶裙，上面系锡绣几何纹、彩线挑花暗纹装饰的前、后围腰，另有青布头帕和绑腿。剑河锡绣是苗族刺绣中的一绝，工艺相当复杂，先用剪刀将锡片剪下长约 18 厘米，宽约 1 毫米的锡条，一头打上小勾，另一头成针头状，再用针头状的锡条穿过家

图2-9-1 剑河南寨苗族锡绣对襟女套装

Picture 2-9-1 Miao's Botton-less Symmetric Woman Outfit in Tin Embroidery at Nanzhai of Jianhe

织布，把打小勾的锡丝条扣住棉线，剪断后将余下的锡条反扣在棉线上，完成扣锡过程。盛装时，女子佩戴百叶银项圈、银花帽、银耳坠、银链锁、银手镯等。锡绣以几何纹为主，既有相同，又有区别，象征着祖先生活的环境和使用的一些工具，如山岭、河谷、屋脊、牛鞍、秤钩、犁耙、弯尺、老人头、小人头等。由于苗族迁徙过程中战事不断，剑河苗族锡绣女装从外观上遗存了古代武士的盔甲之风。

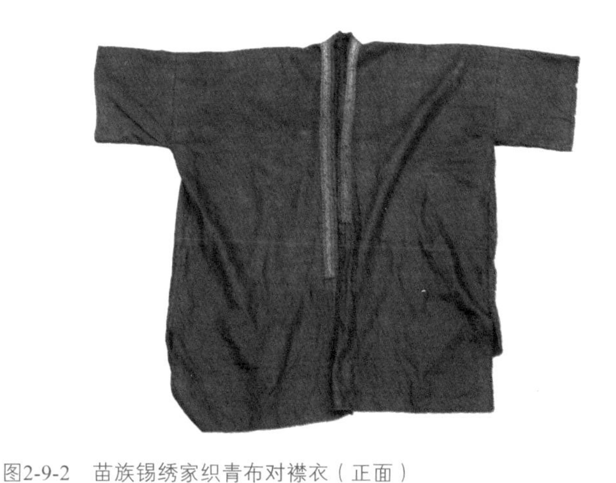

图2-9-2　苗族锡绣家织青布对襟衣（正面）

Picture 2-9-2　Miao's Botton-less Symmetric Outfit Made of Home-woven Black-cloth in Tin Embroidery（Front）

图2-9-3　苗族锡绣家织青布对襟衣（背面）

Picture 2-9-3　Miao's Botton-less Symmetric Outfit Made of Home-woven Black-cloth in Tin Embroidery（Back）

» Name: Miao's Botton-less Symmetric Woman Outfit in Tin Embroidery at Nanzhai of Jianhe

» Time: About 1980

» Collection: W.1741

Represented by the style at towns of Nanzhai and Nanjia, the outfit is popular in towns of Nanzhai, Nanjia, Mingdong, Guanme and Jiupang in Jianhe County of Guizhou Province. The outfit is composed of jacket, pleated skirt, front and back apron and leggings. Made of black cloth, the symmetric jacket is matched by a sash which is decorated with tin embroidered geometric patterns. The black-cloth pleated skirt is tied by front and back aprons which are tin embroidered with geometric patterns and cross-stitched with colorful lines. Tin embroidery of Jianhe is a special skill of all kinds of embroideries, which has extraordinarily complicated technology. Firstly, people should cut out a thin tin about 18 cm of length and 1mm of width, then make a hook at the end of the tin on one side; on the other side, it should be polished like a needle in order to sew easily through the homemade cloth. Then use the hook side of tin to fasten the cotton thread and cut the rest of tin. Women wear some silver ornaments including neck rings, turban, earrings, locks and bracelets at festivals. Tin embroidery are mainly similar and different geometric patterns which symbolize the environment that ancients lived in and tools that ancients used, including mountain range, valley, saddle, steelyard hook, plow harrow, head shape of the old and children. In addition, because of incessant war on the process of migration, the woman outfit of Miao's tin embroidery of Jianhe remains the style of armor of ancient warriors.

图2-9-4　锡绣几何纹背褡

Picture 2-9-4　Sash with Tin Embroidered Geometric Patterns

图2-9-5　锡绣几何纹、彩线挑花暗纹前、后围腰

Picture 2-9-5　Front and Back Apron Tin Embroidered with Geometric Patterns and Cross-stitched with Colorful Threads

图2-9-6 锡绣流苏

Picture 2-9-6 Tin Embroidered Tassels

图2-9-7 青布百褶裙

Picture 2-9-7 Black-cloth Pleated Skirt

» 名称：丹寨雅灰苗族无领对襟百鸟衣（男装）

» 时代：约1980年

» 馆藏：W.1744

主要流行于贵州丹寨、三都、榕江等地，以丹寨雅灰乡送陇为代表。该服饰为当地男子在重大节日活动中的盛装，上衣为无领对襟衫，下为飘带裙，衣裙相连，裙摆下端缀白色羽毛，是服饰中仿生学的巧妙运用。服饰的主色调以绿、黄居多，图案以平绣各种变形蝴蝶纹、鸟纹、龙纹和鱼纹等，主要表达神话传说中对苗族祖先蝴蝶和鹡宇鸟的崇拜，也称为“蝴蝶衣”。每十二年过一次的苗族鼓藏节，男子头戴青布头帕，身穿蝴蝶衣，手拿古瓢琴，与盛装的女子载歌载舞祭奠祖先。其制作工艺繁复，要在蚕吐丝的季节让蚕在门板上吐丝成片，再用开水烫后用小刀片割下来，贴在家织布上，或染成绿色后再贴，所以百鸟衣一般使用蚕丝本色——土黄色或植物染的绿色蚕丝板做底，再进行刺绣。制作一件丹寨雅灰的百鸟衣至少需要两年时间。

» Name: Miao's Collarless Symmetric Man Outfit (Hundred-bird Coat) at Yahui of Danzhai

» Time: About 1980

» Collection: W.1744

Represented by the style at Songlong Village, Yahui Town of Danzhai County, the outfit is distributed in the counties of Danzhai, Sandu and Rongjiang in Guizhou Province. The collarless symmetric jacket is matched by a streamer skirt decorated with white feathers, which is the skillful application of bionics in costume. The festival costume is usually worn at festivals by men and mainly in green and yellow. Many kinds of motifs including butterfly, bird, dragon and fish are flat stitched on the whole outfit, also called Butterfly Outfit, which is the worship for "Butterfly Mom", Miao's ancestor and "Jiyu bird" (the egg hatcher for "Butterfly Mom") in legend. At Guzang Festival every twelve years, the grandest festival of Miao, men wear black headscarf and Butterfly Outfit, hold Gupiao Qin, a kind of traditional Miao's musical instrument, sing and dance with women also dressed in festival costume to worship ancestors. The production process of the hundred-bird coat is complicated. In the season when silkworm spin, the door is prepared for the silkworm to spin into pieces, and then poured by boiling water. After that, the silk plate is cut off and pasted on home-woven cloth or pasted after being dyed into green. Original silk——earth-yellow silk or green silk is usually used as the texture to embroider. It usually takes two years to finish an exquisite suit.

图2-10-1　丹寨雅灰苗族无领对襟百鸟衣（男装背面）

Picture 2-10-1　Miao's Collarless Symmetric Outfit (Hundred-bird Coat) at Yahui of Danzhai (Men's Wear Back)

» 名称：丹寨排莫苗族右衽女套装
» 时代：约1970年
» 馆藏：W.1743

主要流行于贵州丹寨、三都、都匀等地，以丹寨排莫、三都普安为代表。该套装由上衣、长直裙、腰带等组成。上衣以牛血亮布制作，肩、门襟使用平绣、轴线绣和马尾绣绣花草纹、花蝶纹，背部及双袖缝接彩色窝妥纹蜡染花块，腰部微收。下装现为蜡染的两片长直裙，以葵花、石榴、穗子等图案为主，另有彩色腰带作配饰。古装为下图中的马面裙，现在村寨几乎灭迹。盛装时，女子戴三岔银角花、银头花、银项圈，右腰间缀长银链等。

图2-11-1　丹寨排莫苗族右衽女套装（背面）
Picture 2-11-1　Miao's Right Buttoned Woman Outfit at Paimo of Danzhai (Back)

» Name: Miao's Right Buttoned Woman Outfit at Paimo of Danzhai

» Time: About 1970

» Collection: W.1743

Represented by the style at Paimo Village of Danzhai County and Pu'an of Sandu County, the outfit is distributed in the counties of Danzhai, Sandu and Duyun City in Guizhou Province. It is composed of jacket, long straight skirt and sash. Made of bull-blood dyed bright cloth, the jacket is in the fashion of tight-waist and decorated with patterns of flowers, grasses and butterflies in flat embroidery, axis stitch and pony-tail stitch on shoulders and lapels, and colorful batik vortex-pattern pieces stitched on back and sleeves. Nowadays, the long straight skirt is made of two pieces of batik cloth with hand-drawn patterns of sunflower, guava and tassel and decorated with colorful sash. The picture below "the skirt with horse face" is the ancient costume which is almost extinct at the village now. At festivals, women adorn themselves with silver ornaments including horn-like headdress flowers, headdress flowers and neck rings, and decorate with long silver chain at right waist.

图2-11-2　后背蜡染窝妥纹

Picture 2-11-2　Batik Vortex Patterns on Back

图2-11-3　衣袖窝妥纹和花纹

Picture 2-11-3　Vortex Patterns and Flower Patterns on Sleeves

图2-11-4　裙身蜡染葵花、石榴、穗子图案

Picture 2-11-4　Batik Patterns of Sunflower, Guava and Tassel on Skirt

图2-11-5 平绣、轴线绣、马尾绣的花蝶

Picture 2-11-5 Butterfly Patterns in Flat Embroidery, Axis Stitch and Pony-tail Stitch

图2-11-6 蜡染花纹

Picture 2-11-6 Batik Flower Pattern

图2-11-7　丹寨苗族女盛装蜡染盛装（黄晓海　摄）

Picture 2-11-7　Miao's Woman Batik Festival Costume at Danzhai (Photographed by Huang Xiaohai)

» 名称：榕江兴华摆贝苗族无领对襟百鸟衣（男装）
» 时代：约1980年
» 馆藏：W.1746

主要流行于贵州榕江、丹寨、三都等地，以榕江兴华摆贝为代表。该服饰为当地男子在鼓藏节中的盛装，上为衣，下为飘带裙，衣裙相连，裙摆下端缀白色羽毛，服饰以织锦几何纹为主，少量的鸟纹等，呈左右或上下对称图形，色彩以绿色为基调，上衣与飘带裙连为一体的造型样式，保留了春秋时期“深衣”的款式特征。当地盛行每十二年过一次的鼓藏节，也是苗族祭祖的重要盛典，每个村寨的鼓藏节时间也不同。节日期间，鼓藏头们身穿鼓藏服，头戴银抹额，每家每户挂鼓藏帆，进行隆重的祭祀活动。邻村的村民也会借此机会盛装来走访亲友，一起过节。举办盛大的鼓藏节也是苗族对祖先的崇敬和缅怀。织锦工艺的鼓藏服织造复杂，需要至少两三年左右才能完成。

图2-12-1　榕江兴华摆贝苗族无领对襟百鸟衣（男装）
Picture 2-12-1　Miao's Collarless Symmetric Outfit (Hundred-bird Coat) at Baibei, Xinghua of Rongjiang (Men's)

» Name: Miao's Collarless Symmetric Man Outfit (Hundred-bird Coat) at Baibei, Xinghua of Rongjiang

» Time: About 1980

» Collection: W.1746

Represented by the style at Baibei Village, Xinghua Town of Rongjiang County, the outfit is distributed in the counties of Rongjiang, Danzhai and Sandu in Guizhou Province. The collarless symmetric jacket is connected with a streamer skirt decorated with white feathers. The festival costume is usually worn at Guzang Festival by men and mainly in green. The outfit is decorated with brocaded geometric patterns and bird patterns which are symmetric up to down or left to right. The connected suit style remains the features of "Shenyi" of Chunqiu Period. Guzang Festival every twelve years is the grandest festival of Miao for worshiping ancestors. The date of Guzang Festival is different in different villages. During the festival, the heads of Guzang wear Guzang costume and silver headband. Every house is decorated with Guzang bag and Miao people do grand worshiping activities. People from neighboring villages will visit relatives during the time to celebrate the festival. It usually takes two to three years to finish an exquisite suit.

» 名称：榕江兴华摆贝苗族无领对襟百鸟衣（女装）
» 时代：约1980年
» 馆藏：W.1745

主要流行于贵州榕江、丹寨、三都等地，以榕江兴华摆贝为代表。该服饰为当地女子在鼓藏节中的盛装，由上衣、飘带裙、三角围腰、织锦腰带和亮布脚笼组成。上装为亮布衣，衣肩和衣襟配有几何纹彩色织锦。下装为彩色织锦飘带裙，裙摆底缀白色羽毛。三角形围腰由三块彩色织锦拼接，织锦图案多为几何纹，色泽饱满。另配有亮布脚笼和织锦花腰带。每十二年一次的鼓藏节，男女身穿百鸟衣祭祖，因此也称“鼓藏服”，羽毛的使用据说与苗族祖先的捕猎习俗有关，鸟伴随这支迁徙的民族，也是苗族的吉祥物，故在飘带裙底缀羽毛。节日期间，女子戴银头花、银梳、银项圈、银耳环等。织锦百鸟服以对称的几何纹为主，工艺繁复，每套服饰至少需要两至三年才能完成。

图2-13-1 榕江兴华摆贝苗族无领对襟百鸟衣（女装）正面

Picture 2-13-1 Miao's Collarless Symmetric Woman Outfit (Hundred-bird Coat) at Baibei, Xinghua of Rongjiang (Front)

图2-13-2 榕江兴华摆贝苗族无领对襟百鸟衣（女装）背面

Picture 2-13-2 Miao's Collarless Symmetric Woman Outfit (Hundred-bird Coat) at Baibei, Xinghua of Rongjiang (Back)

» Name: Miao's Collarless Symmetric Woman Outfit (Hundred-bird Coat) at Baibei, Xinghua of Rongjiang

» Time: About 1980

» Collection: W.1745

Represented by the style at Baibei Village, Xinghua Town of Rongjiang County, the outfit is distributed in the counties of Rongjiang, Danzhai and Sandu in Guizhou Province. Composed of jacket, streamers, triangular apron, brocaded sash and bright-cloth leggings, this festival costume is worn by local Miao women at Guzang Festival. Made of bright-cloth, the jacket is decorated with geometric colorful brocade on shoulders and lapels. The brocaded streamer skirt is decorated with white feathers and a triangular apron composed of three-piece geometric colorful brocade. At Guzang Festival every twelve years, men and women wear hundred-bird coat to worship ancestors. As lucky symbols, feathers are used in costume producing because they are related with hunting custom and migration of Miao's ancestors. During the festival, women wear silver ornaments including headdress flowers, comb, neck-rings and earrings. It usually takes two to three years to finish an exquisite suit.

图2-13-3 彩色挑绣几何纹样亮布衣襟局部图

Picture 2-13-3 Colorful Geometric Patterns Cross-stitched on Bright-cloth Lapel (Section)

图2-13-4　彩色挑绣几何纹样亮布衣袖局部图

Picture 2-13-4　Colorful Geometric Patterns Cross-stitched on Bright-cloth Sleeves (Section)

图2-13-5　挑绣几何纹三角围腰局部图

Picture 2-13-5　Cross-stitched Geometric Triangular Apron (Section)

图2-13-6　挑绣几何纹样缀白色鸡毛飘带裙局部图

Picture 2-13-6　Cross-stitched Geometric Patterns on Streamer Skirt with White Feathers (Section)

» 名称：榕江兴华摆贝苗族无领对襟百鸟衣
» 时代：约1960年
» 馆藏：W.1747

主要流行于贵州榕江、丹寨、三都等地，以榕江兴华摆贝为代表。该服饰为当地女子在鼓藏节等重大活动中的盛装，上为衣，下为飘带裙，衣裙相连，裙摆底缀白色羽毛。该裙以绿色绸缎布为底，不是蚕丝板，也不是织锦工艺，主要运用平绣针法形成衣饰左右对称图案，制作这样的百鸟服相对简单。服饰以传统的鸟纹为主，配有花、虫、鱼、蝴蝶等图案，主要表达苗族对"蝴蝶妈妈"故事中鹡宇鸟的崇拜，故称"百鸟衣"。节日期间，女子戴银头花、银梳、银项圈、银耳环等。制作这样的百鸟衣需要约一年的时间。

图2-14-1　榕江兴华摆贝苗族无领对襟百鸟衣

Picture 2-14-1　Miao's Collarless Symmetric Outfit (Hundred-bird Coat) at Baibei, Xinghua of Rongjiang

图2-14-2 衣袖花纹

Picture 2-14-2 Flower Patterns on Sleeves

图2-14-3 飘带裙中几何纹

Picture 2-14-3 Geometric Patterns on Streamers

» Name: Miao's Collarless Symmetric Outfit (Hundred-bird Coat) at Baibei, Xinghua of Rongjiang

» Time: About 1960

» Collection: W.1747

Represented by the style at Baibei Village, Xinghua Town of Rongjiang County, the outfit is distributed in the counties of Rongjiang, Danzhai and Sandu in Guizhou Province. Composed of jacket and streamers with white feathers, this festival costume is worn by local Miao women at Guzang Festival. Made of green brocade, the skirt is flat stitched with symmetric patterns. The technique of the costume producing is comparatively simple. The patterns on the costume mainly are birds, decorated with patterns of flower, insect, fish and butterfly. Therefore, the costume is also called "Hundred-bird coat" to worship "Jiyu bird" (the egg hatcher for "Butterfly Mom") in the story of "Butterfly Mom". During the festival, women wear silver ornaments including headdress flowers, comb, neck-rings and earrings. It usually takes one year to finish the costume.

图2-14-4 后背鸟纹和蝴蝶纹

Picture 2-14-4 Patterns of Bird and Butterfly on Back

图2-14-5 后背花纹

Picture 2-14-5 Flower Patterns on Back

» 名称：雷山桃江苗族无领对襟女套装
» 时代：约1970年
» 馆藏：W.1751

主要流行于贵州雷山、丹寨等地，以雷山桃江为代表。该套装由上衣、百褶裙、飘带裙、围腰、腰带等组成。上衣为敞袖无领对襟衣，双袖、前胸以挑花菱形纹、方块纹等为主，象征祖先耕种的农田。下装为双层裙，里裙为七寸的青布百褶裙，外裙为绿色织锦带麦穗飘带裙穿在身后。腰间系多层几何纹样围腰，外加织锦带穗

图2-15-1　雷山桃江苗族无领对襟女套装
Picture 2-15-1　Miao's Collarless Symmetric Woman Outfit at Taojiang of Leishan

宽腰带。该支系的青布百褶裙只有七寸长，故又称为短裙苗，穿着时以多为美，一般要穿五到六条百褶裙，织锦飘带裙在百褶裙外呈隆起状，若美丽的锦鸡，也象征丰臀，子孙繁衍的盛世。盛装时，女子戴银花帽、饰满银泡的围腰、银梳、银项圈、银手镯，衣服上饰银片。

图2-15-2　挑花菱形纹、方块纹

Picture 2-15-2　Patterns of diamond and Square in Cross-stitch

图2-15-3　几何纹、花纹围腰

Picture 2-15-3　Apron with Geometric Patterns and Flower Patterns

» Name: Miao's Collarless Symmetric Woman Outfit at Taojiang of Leishan

» Time: About 1970

» Collection: W.1751

Represented by the style at Taojiang Town of Leishan County, the outfit is distributed in the counties of Leishan and Danzhai in Guizhou Province. It is composed of jacket, pleated skirt, streamers, apron and sash. With loose sleeves, the collarless jacket is decorated with cross-stitched patterns of diamond and square which symbolize the fields ancestors plowed. The double-layer skirt is composed with an inner black-cloth pleated skirt and outer green brocaded streamers tied at back. The skirt is decorated with an apron with multi-layer geometric patterns and wide brocaded sash. The skirt is only 7 inches long, so the branch of Miao who wear such skirt is also called "Miao in Mini-skirt". Wearing five to six pleated skirts to raise streamers, women are like beautiful golden pheasants, which represents offspring multiplication with full buttocks. Wearing the jacket decorated with silver pieces, women adorn themselves with silver ornaments including turban, comb, neck-rings and bracelets and aprons decorated with silver bubbles at festivals.

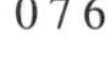

图2-15-4 织锦带穗宽腰带局部图

Picture 2-15-4 Wide Brocaded Sash (Section)

图2-15-5 青布百褶裙局部图

Picture 2-15-5 Black-cloth Pleated Skirt (Section)

» 名称：黎平水口侗族镂空对襟女套装
» 时代：约1980年
» 馆藏：W.1750

主要流行于贵州黎平水口镇及周边村寨，以水口镇为代表。该套装由上衣、百褶裙、胸兜、脚笼等组成。上衣为家织青布无领、无扣对襟衣，七分衣袖，前、后三分之二镂空编织，下缀鹅毛杆和料珠。下装为猪（牛）血百褶裙。另有绣花胸兜和青布脚笼，为女子夏天便装。黎平水口镇位于贵州省东南部，南邻广西壮族自治区三江侗族自治县，该地区夏季气候炎热，女子流行穿易于散热的镂空衣。用猪（牛）血浆制的亮布百褶裙坠感较好，制作颇具特色，褶皱间刷上厚厚的猪（牛）血混合物，通过反复蒸煮和晾晒，使混合物牢牢地凝固在褶皱上形成亮片状，太阳下闪闪发光。

图2-16-1 黎平水口侗族镂空对襟女套装（背面）

Picture 2-16-1 Dong’s Hollowed-out Symmetric Woman Outfit at Shuikou of Liping (Back)

» Name: Dong's Hollowed-out Symmetric Woman Outfit at Shuikou of Liping

» Time: About 1980

» Collection: W.1750

Represented by the style at Shuikou Town, the outfit is popular in Shuikou Town and its surrounding villages of Liping County in Guizhou Province. With three-quarter sleeves, two-thirds hollowing-out woven body, and decorated with goose feather rods and beads, the collarless unbuttoned symmetric jacket is made of home-woven black-cloth. The pleated skirt is dyed with blood of bulls and pigs. Composed with jacket, pleated skirt, embroidered under-apron and black-cloth leggings, the outfit is flat clothes worn by women in summer. Located in the south-east of Guizhou Province and adjacent to Sanjiang Dong Autonomous County of Guangxi Zhuang Autonomous Region in the south, Shuikou Town of Liping County is very hot in summer so that the hollowed-out outfit is popular. The pleated skirt is made of bright-cloth and has a better hanging nature because it is dyed with thick mixture of pig and bull blood and stewed and dried repeatedly. This technique is used to help freeze mixture on the pleated skirt to keep the pleat permanently and shine under sunshine.

图2-16-2 家织青布

Picture 2-16-2 Home-woven Black-cloth

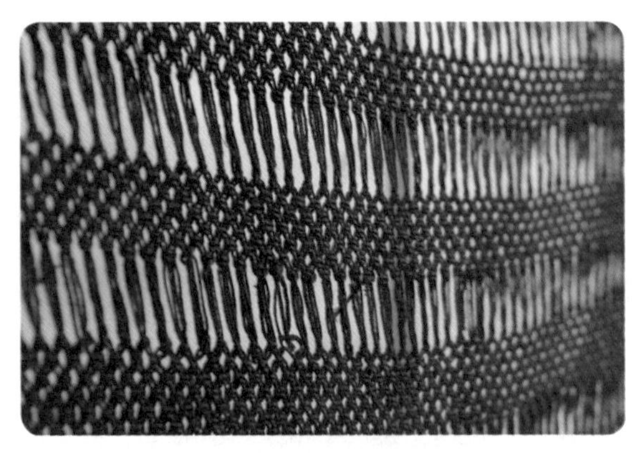

图2-16-3　镂空编织

Picture 2-16-3　Hollowed-out Weaving

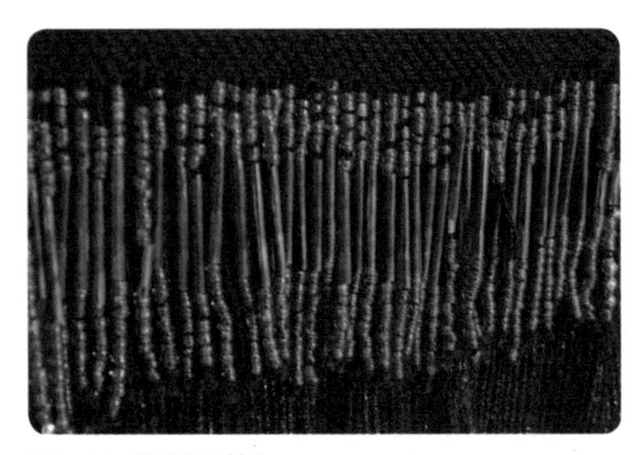

图2-16-4　鹅毛杆和料珠

Picture 2-16-4　Goose Feather Rod and Bead

图2-16-5　猪（牛）血亮布百褶裙

Picture 2-16-5　Bright-cloth Pleated Skirt Dyed with Blood of Pigs and Bulls

» 名称：黎平尚重侗族无领右衽大襟女套装
» 时代：约1960年
» 馆藏：W.1755

主要流行在贵州黎平、从江、榕江等地，以黎平尚重镇为代表。该套装由上衣、百褶裙、围腰、霞帔、绑腿等组成。上衣为紫色缎面无领右衽大襟衣，沿衣襟、双袖边以红色布为底，用轴线绣、绞绣绣制二龙抢宝、缠枝花纹，并缝钉许多亮片和塑料珠为饰。下装为青布百褶裙，腰前系蓝色缎面方形围腰，围腰上段红色布为底轴线绣、绞绣二龙抢宝图案；下端四角贴布绣云卷纹，四周轴线绣、绞绣缠枝花纹，配色和谐俏丽。盛装时，女子配有贴布云卷纹，轴线绣、绞绣缠枝花纹带穗子霞帔，另戴银头花、银项圈、银手镯等。

图2-17-1　黎平尚重侗族无领右衽大襟女套装（正面）

Picture 2-17-1　Dong's Collarless Right-buttoned Woman Outfit at Shangzhong of Liping (Front)

图2-17-2　黎平尚重侗族无领右衽大襟女套装（背面）

Picture 2-17-2　Dong's Collarless Right-buttoned Woman Outfit at Shangzhong of Liping (Back)

» Name: Dong's Collarless Right-buttoned Woman Outfit at Shangzhong of Liping

» Time: About 1960

» Collection: W.1755

Represented by the style at Shangzhong Town of Liping, the outfit is popular in counties of Liping, Congjiang and Rongjiang in Guizhou Province. It is composed of jacket, pleated skirt, apron, cape and leggings. The jacket is a collarless right-buttoned purple brocaded coat with patterns of "two dragons striving for pearl" and "winding stem" axis stitched and twisting stitched on the rims of the lapel and sleeves in red cloth and decorated with blinging pieces and plastic beads. The pleated skirt is tied by a square blue brocaded apron which is decorated with patterns of "two dragons striving for pearl" axis stitched and twisting stitched on the upper part in red cloth and applique stitched cirrus cloud pattern at the four corners surrounded by "winding stem" pattern. Adorned with silver ornaments including headdress flowers, neck-rings and bracelets, women wear capes which are pasting stitched with cirrus cloud pattern and axis stitched and twisting stitched "winding stem" patterns at festivals.

图2-17-3　轴线绣、绞绣花纹

Picture 2-17-3　Patterns in Axis Stitch and Twisting Stitch

图2-17-4　轴线绣、绞绣缠枝花纹

Picture 2-17-4　“Winding Stem” Patterns in Axis Stitch and Twisting Stitch

图2-17-5　贴布绣云卷纹

Picture 2-17-5　Cirrus Cloud Patterns in Applique Stitch

图2-17-6 围腰整体图

Picture 2-17-6 Apron (Overall)

» 名称：榕江平江滚仲苗族贴布绣对襟女套装
» 时代：约1970年
» 馆藏：W.1754

主要流行于贵州榕江周边等地，以榕江平江滚仲为代表。该套装由头帕、上衣、胸兜、百褶裙、围腰、脚笼和脚带组成。上衣为贴布绣龙纹、云纹亮布衣。胸兜呈菱形，云雷纹蜡染拼缝，上以红、绿轴段方块点缀，胸前平绣折枝花。下装为雷纹、花草纹蜡染布、青布及彩色绸缎相间百褶裙。长方形围腰周边平绣龙纹、团花纹、草纹。另有几何纹挑花脚笼和织锦脚带。服饰中的云雷纹表现了苗族祖先在远古时期对自然界的崇拜。盛装时，女子戴银项圈、银链条等。

图2-18-1　榕江平江滚仲苗族贴布绣对襟女套装（正面）
Picture 2-18-1　Miao’s Symmetric Woman Outfit in Applique Stitch at Gunzhong, Pingjiang of Rongjiang (Front)

» Name: Miao's Symmetric Woman Outfit in Applique Stitch at Gunzhong, Pingjiang of Rongjiang

» Time: About 1970

» Collection: W.1754

Represented by the style at Gunzhong, Pingjiang Town of Rongjiang, the outfit is popular in areas of Rongjiang County in Guizhou Province. It is composed of headscarf, jacket, under-apron, pleated skirt, apron, leggings and straps. The jacket is a bright coat with patterns of dragon and cloud in applique stitch. In the shape of diamond and decorated with batik stitched patterns of cloud and thunder, the under-apron is ornamented with red and green squares and flat stitched "zhezhi flower". The pleated skirt is alternated by batik, black-cloth and colored brocade with patterns of thunder, flower and grass. The rectangular apron is flat embroidered with patterns of dragon, flower and grass around. In addition, women wear leggings with cross-stitched geometric patterns and brocaded straps. The patterns of cloud and thunder represent the Miao's worship for nature in antient times. Women wear silver neck-rings and silver chains at festivals.

图2-18-2　榕江平江滚仲苗族贴布绣对襟女套装（背面）

Picture 2-18-2　Miao's Symmetric Woman Outfit in Applique Stitch at Gunzhong, Pingjiang of Rongjiang (Back)

图2-18-3　织锦头帕

Picture 2-18-3　Brocaded Headscarf

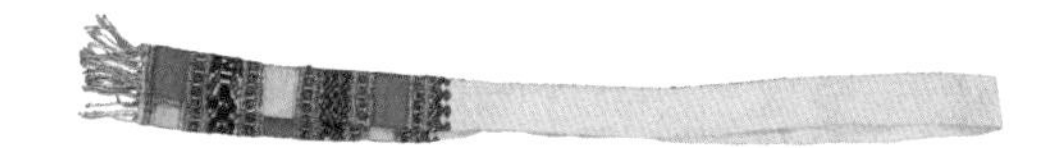

图2-18-4　织锦脚带

Picture 2-18-4　Brocaded Straps

图2-18-5　贴布绣围腰

Picture 2-18-5　Apron in Applique Stitch

图2-18-6　贴布绣龙纹、云纹、花纹

Picture 2-18-6　Patterns of Dragon, Cloud and Flower in Applique Stitch

图2-18-7　胸兜蜡染几何纹和花纹

Picture 2-18-7　Under-apron with Batik Geometric Patterns and Flower Patterns

图2-18-8　蜡染雷纹、花纹和彩色绸缎相间的百褶裙

Picture 2-18-8　Pleated Skirt with Batik Patterns of Thunder and Flower and Alternated by Colored Brocade

» 名称：榕江仁里苗族轴线绣无领右衽大襟女套装
» 时代：约1990年
» 馆藏：W.1758

主要流行于贵州榕江仁里乡及周边村寨，以仁里乡加榜、龚街为代表。该套装由上衣、百褶裙、围腰、绑腿等组成。上衣为家织青布无领右衽大襟衣，双袖、衣襟上轴线绣龙纹、鱼纹、鸟纹、蝶纹和缠枝纹等，栏杆花边饰之。下装为青布百褶裙，外系青布轴线绣双鱼跳龙门、凤凰朝阳等吉祥图案围腰，亮片散钉于刺绣间。缠青布绑腿，系织锦花带。盛装时，女子头插银头花，佩戴银项圈、银颈链、银手镯、银耳坠等。

图2-19-1　榕江仁里苗族轴线绣无领右衽大襟女套装

Picture 2-19-1　Miao's Collarless Right-buttoned Woman Outfit at Renli of Rongjiang

» Name: Miao's Collarless Right-buttoned Woman Outfit at Renli of Rongjiang

» Time: About 1990

» Collection: W.1758

Represented by the style at Jiabang and Gongjie of Renli Village, the outfit is popular in Renli Village and its surrounding areas in Guizhou Province. It is composed of jacket, pleated skirt, apron and leggings. The jacket is a collarless right-buttoned black-cloth coat with patterns of dragon, fish, bird, butterfly and tangled branch axis stitched on the hem of lapel and sleeves and decorated with lace. The pleated skirt is tied by a black-cloth apron which is axis stitched with patterns of "two fish jumping over the Dragon Gate" and "phoenix facing sun" and decorated with blinging pieces. In addition, women wear black-cloth leggings and brocaded straps. Women wear silver ornaments including headdress flowers, neck-rings, chains, bracelets and earrings at festivals.

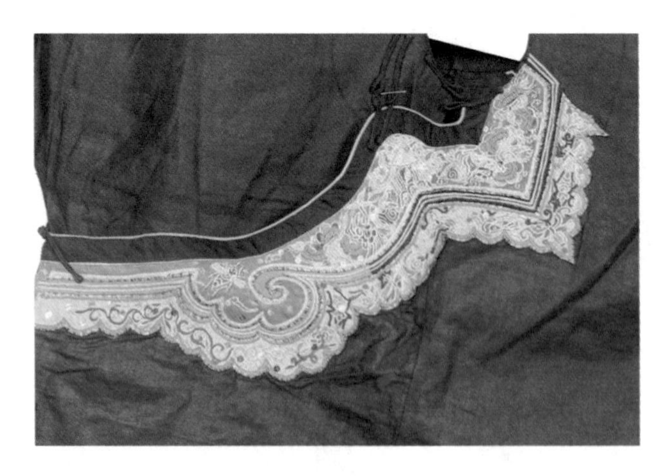

图2-19-2　衣襟贴布轴线绣花纹

Picture 2-19-2　Flower Patterns Axis Stitched on the Hem of Lapel

图2-19-3　衣袖轴线绣鸟纹、蝶纹和平绣花纹

Picture 2-19-3　Patterns of Bird, Butterfly and Flower Axis Stitched on Sleeves

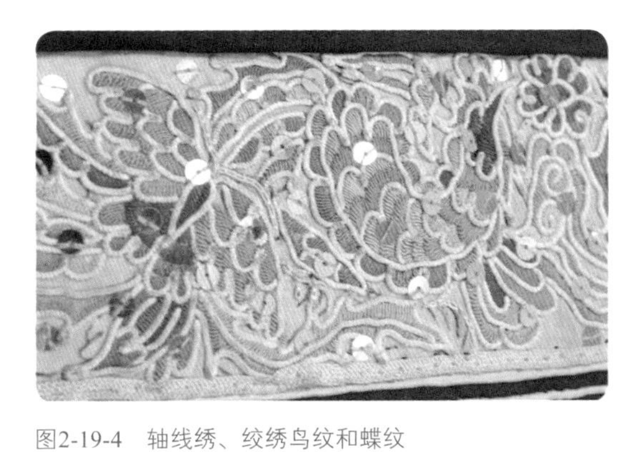

图2-19-4　轴线绣、绞绣鸟纹和蝶纹

Picture 2-19-4　Patterns of Bird and Butterfly in Axis Stitch and Twisting Stitch

图2-19-5　平绣花纹

Picture 2-19-5　Flower Patterns in Flat Stitch

» 名称：从江丙妹岜沙苗族无领左衽男套装
» 时代：约2000年
» 馆藏：W.1757

主要流行于贵州从江丙妹周边等地，以从江丙妹岜沙为代表。该套装由上衣、大脚裤、挎包、腰带等组成。上衣为家织青布无领左衽衫，下装为家织青布阔腿裤。农历 11 月 19 日的芦笙节是岜沙最隆重的节日，他们以家族为单位盛装出席，还可邀请其他村寨的亲朋好友盛装一起过节，载歌载舞举行隆重的砍牛祭祖活动。进芦笙场后人们面向东方倒退着往山坡上走，据说这是怀念祖先从东方逐渐退居到高山的意思，每吹完一首芦笙曲都要集体鸣枪，场面极为热烈壮观。盛装时，男子斜跨锁绣羽绒流苏包，腰系锁绣羽绒流苏彩色腰带，腰带中部锁绣锦鸡羽毛图案，塑料珠饰之。相传岜沙的祖先迁徙途中以狩猎为生，因此妇女的胸兜、男子的背心和背包上均饰有与鸟的绒毛相关的刺绣纹样。岜沙崇尚树文化，枫树是他们的生命树，其头饰尤其特别，从小在头顶蓄发，将周围的头发剃去，被称为“户棍”。

图2-20-1　从江丙妹岜沙苗族无领左衽男套装（正面）

Picture 2-20-1　Miao’s Collarless Left-buttoned Man Outfit at Basha, Bingmei of Congjiang (Front)

图2-20-2　从江丙妹邑沙苗族无领左衽男套装（背面）
Picture 2-20-2　Miao's Collarless Left-buttoned Man Outfit at Basha, Bingmei of Congjiang (Back)

图2-20-3　家织青布
Picture 2-20-3　Home-woven Black-cloth

» Name: Miao's Collarless Left-buttoned Man Outfit at Basha, Bingmei of Congjiang

» Time: About 2000

» Collection: W.1757

Represented by the style at Basha, Bingmei Village of Congjiang County, the outfit is popular in Bingmei Village and its surrounding areas in Guizhou Province. It is composed of jacket, loose pants, satchel and sash. Made of black-cloth, the jacket is a collarless left-buttoned coat and matched with home-woven loose pants. Lusheng Festival on Nov. 19th lunar calendar is the grandest festival in Basha. Dressed in their holiday best, each family invites their relatives and friends from other villages to attend the ceremonious sacrificing activities with dancing and singing. After entering Lusheng field, facing the east, people walk back to the hillside, which is said to memorize ancestors retreating to the mountains from the East. It is very magnificent that people fire rifles into the air together after finishing a piece of Lusheng music. Tied with colored sashes adorned with stitched "down flower" patterns and tassels, feather patterns and plastic beads, men wear satchels decorated with feather tassels in their festival costumes. In legend, ancestors of Basha hunted for living during migration, so the patterns with feathers and downs are embroidered on women's under-aprons and men's vests and satchels. People from Basha worship tree culture and maples are their life trees. The headwear of men is very special. Men wear small topknots with surrounding hair shaved, which is called "hugun".

图2-20-4 锁绣羽绒花流苏彩色腰带局部图

Picture 2-20-4 Colored Sashes Adorned with Pin Embroidered Down Flower Patterns and Tassels

图2-20-5 锁绣鱼鸟纹

Picture 2-20-5 Pin Embroidered Patterns of Fish and Bird

图2-20-6 锁绣羽绒花纹

Picture 2-20-6 Pin Embroidered Down Flowers

图2-20-7　芦笙节

Picture 2-20-7　Lusheng Festival

» 名称：从江丙妹岜沙苗族对襟女套装
» 时代：约1980年
» 馆藏：W.1753

主要流行于贵州从江丙妹周边等地，以从江丙妹岜沙为代表。该套装由上衣、百褶裙、胸兜、脚笼和脚带组成。上衣为对襟无领紧身衣，袖口、衣摆饰彩色挑花几何纹，三角形胸兜用亮布制作，上有锁绣羽绒花。下装为枫香染百褶裙。另配有亮布挑花脚笼。相传岜沙祖先迁徙途中以狩猎为生，妇女的胸兜、男子的背心和背包上均饰有与鸟的绒毛相关的刺绣纹样。盛装时，女子戴银项圈、银耳环等。

图2-21-1 从江丙妹岜沙苗族对襟女套装（正面）
Picture 2-21-1 Miao's Symmetric Woman Outfit at Basha, Bingmei of Congjiang (Front)

图2-21-2　从江丙妹岜沙苗族对襟女套装（背面）

Picture 2-21-2　Miao's Symmetric Woman Outfit at Basha, Bingmei of Congjiang (Back)

» Name: Miao's Symmetric Woman Outfit at Basha, Bingmei of Congjiang

» Time: About 1980

» Collection: W.1753

Represented by the style at Basha, Bingmei Village of Congjiang County, the outfit is popular in Bingmei Village and its surrounding areas in Guizhou Province. It is composed of jacket, pleated skirt, under-apron, leggings and straps. The collarless symmetric tight jacket is cross-stitched with colored geometric patterns on sleeves and the hem of jacket. Made of bright cloth, the under-apron is pin stitched with "down flower" patterns. The pleated skirt dycd with maple is matched with embroidered bright-cloth leggings. According to legend, ancestors of Basha hunted for living during migration, so the patterns with feathers and downs are embroidered on women's under-aprons and men's vests and satchels. At festivals, women wear silver neck-rings and silver earrings.

图2-21-3　挑花几何纹

Picture 2-21-3　Cross-stitched Geometric Patterns

图2-21-4 锁绣羽绒花纹和鱼纹

Picture 2-21-4 Patterns of "Down Fower" and Fish in Pin Embroidery

图2-21-5 枫香染百褶裙

Picture 2-21-5 Pleated Skirt Dyed with Maple

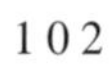

图2-21-6 亮布挑花脚笼

Picture 2-21-6 Bright-cloth Leggings in Cross-stitch

图2-21-7 从江岜沙苗族女盛装

Picture 2-21-7 Miao's Woman Festival Costume at Basha of Congjiang

» 名称：榕江古州苗族无领左衽男套装

» 时代：约2000年

» 馆藏：W.1756

主要流行于贵州榕江、从江等地，以榕江古州为代表。该套装由上衣、大脚裤、腰带和头帕等组成。上衣为无领左衽亮布衫，门襟平绣多色花块，间钉亮片装饰，色彩鲜艳，袖口处以彩色布条装饰，下装为亮布大脚裤，裤脚边平绣各色花块，腰间系红、绿、紫三色挑绣彩带，另有亮布头帕。榕江古州苗族先祖迁徙中曾以水稻种植和狩猎为生，定居榕江后，由于气候较热，在服饰的制作上有意将裤管裁剪得较大，目的是易于奔跑和散热。

图2-22-1　榕江古州苗族无领左衽男套装（正面）

Picture 2-22-1　Miao’s Collarless Left-buttoned Man Outfit at Guzhou of Rongjiang (Front)

» Name: Miao's Collarless Left-buttoned Man Outfit at Guzhou of Rongjiang
» Time: About 2000
» Collection: W.1756

Represented by the style at Guzhou Villiage of Rongjiang County, the outfit is popular in counties of Rongjiang and Congjiang in Guizhou Province. It is composed of jacket, loose pants, sash and headscarf. Made of bright cloth, the collarless left-buttoned jacket is flat stitched with multi colored pieces adorned with blinging pieces and colored straps at the cuff of sleeves. The loose bright-cloth pants are flat stitched with pieces in different colors at the hem of pants and tied with colored straps in red, green and purple. In addition, men wear bright-cloth headscarf. Miao ancestors at Guzhou of Rongjiang lived on rice cultivation and hunting in migration. After settling down in Rongjiang County, they tailored their pants loose to cool themselves because it was quite hot there.

图2-22-2 榕江古州苗族无领左衽男套装（背面）
Picture 2-22-2 Miao's Collarless Left-buttoned Man Outfit at Guzhou of Rongjiang (Back)

图2-22-3　红、绿、紫三色挑花彩带

Picture 2-22-3　Cross-stitched Colored Straps in Red, Green and Purple

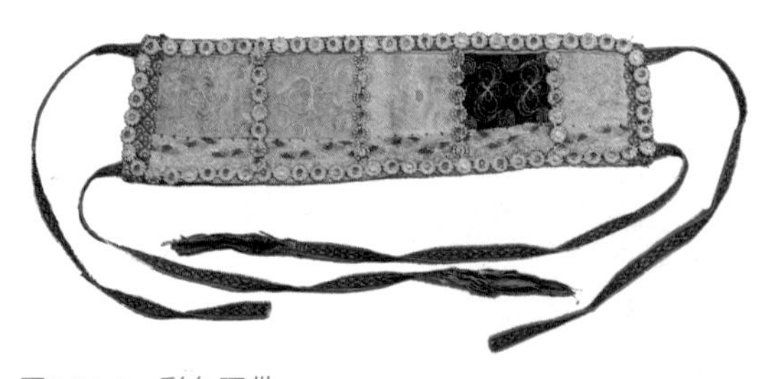

图2-22-4　彩色腰带

Picture 2-22-4　Colored Sash

图2-22-5　亮布头帕

Picture 2-22-5　Bright-cloth Headscarf

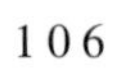

图2-22-6 平绣几何纹和花纹

Picture 2-22-6 Geometric Patterns and Fower patterns in Flat Stitch

图2-22-7 平绣几何纹

Picture 2-22-7 Flat Stitched Geometric Patterns

» 名称：榕江古州苗族无领右衽女套装
» 时代：约1990年
» 馆藏：W.1752

主要流行于贵州榕江、从江等地，以榕江古州为代表。该套装由上衣、百褶裙、围腰和脚笼组成。上衣为右衽亮布衫，后腰及两侧有以平绣、轴线绣为主的绣片，其间钉亮片装饰色彩华丽，且与裙同长，下装为青布百褶素裙，另有数纱绣板凳纹围腰和亮布脚笼。该地区气候炎热，人们喜好超薄的家织布做的亮布为服饰的主材，其透气效果好。盛装时，女子戴银锁、银簪子、银项圈、银手镯等。

图2-23-1　榕江古州苗族无领右衽女套装
Picture 2-23-1　Miao's Collarless Right-buttoned Woman Outfit at Guzhou of Rongjiang

» Name: Miao's Collarless Right-buttoned Woman Outfit at Guzhou of Rongjiang

» Time: About 1990

» Collection: W.1752

Represented by the style at Guzhou Villiage of Rongjiang County, the outfit is popular in counties of Rongjiang and Congjiang in Guizhou Province. It is composed of jacket, pleated skirt, apron and leggings. Made of bright cloth, the right-buttoned jacket is flat stitched and axis stitched with embroidery pieces on back waist and sides and adorned with blinging pieces. The black-cloth pleated skirt is matched with apron yarn-counting stitched patterns of bench and bright-cloth leggings. People prefer to use home-woven bright cloth to make costumes to cool themselves because it is quite hot there. Women wear silver ornaments including locks, hairpins, neck-rings and bracelets at festivals.

图2-23-2　榕江古州苗族女盛装

Picture 2-23-2　Miao's Woman Festival Costume at Guzhou of Rongjiang

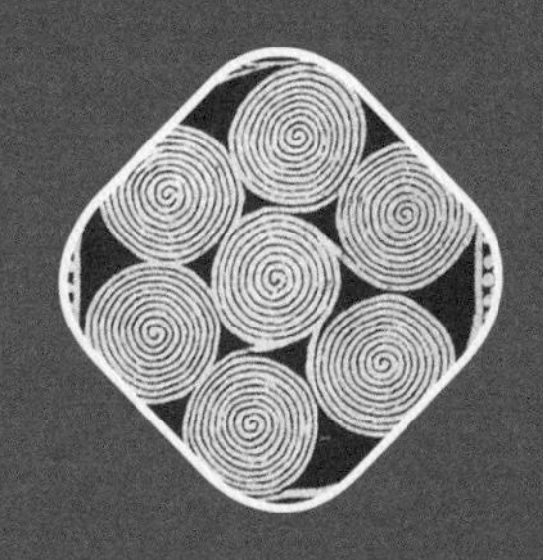

第三章
安顺地区少数民族服饰

Chapter 3 Ethnic Group's Costumes and Ornaments in Anshun Area

» 名称：安顺黑石头寨苗族对襟女套装
» 时代：约1960年
» 馆藏：W.1766

主要流行于贵州安顺、普定等地，以安顺黑石头寨为代表。该套装由上衣、直筒裹裙、围腰、腰带等组成。上衣为齐腰红色对襟衣，双袖、前胸、后背、门襟、衣下摆分别刺绣变形的鸟纹、花草纹、火焰纹等。下装为宽幅直筒裙，裙腰红底刺绣火焰纹、几何纹。另有白布镶蜡染的围腰和织锦或刺绣腰带。传说此套装人的一生只穿两次，第一次是出嫁时；第二次是去世入棺时，要穿这样的盛装才能找到他们的祖宗，他们的灵魂才有安放之处。后演变为凡有喜事即可穿。盛装时，以银链饰发。

图3-1-1 安顺黑石头寨苗族对襟女上衣
Picture 3-1-1 Miao's Symmetric Woman Jacket at Black Stone Village of Anshun Area

» Name: Miao's Symmetric Woman Outfit at Black Stone Village of Anshun Area

» Time: About 1960

» Collection: W.1766

Represented by the style at Black Stone Village of Anshun City, the outfit is popular in Anshun City and Puding County in Guizhou Province. It is composed of jacket, straight skirt, apron and sash. The red waist-length symmetric jacket is embroidered with transformed patterns of bird, flower and grass and flame on sleeves, breast, back, lapel and hems. The straight pleated skirt is embroidered with geometric patterns and flame patterns in red base on the waist of the skirt. In addition, women wear white-cloth and batik aprons and brocaded or embroidered sash. It is said that the outfit is only worn twice in lifetime, at wedding ceremony and at funeral. People in their holiday best can find their ancestors and settle their spirit after death. Nowadays, women are dressed with silver chains ornamented on hair at festivals to celebrate.

图3-1-2　安顺黑石头寨苗族对襟女套装（背面）

Picture 3-1-2　Miao's Symmetric Woman Outfit at Black Stone Village of Anshun Area (Back)

图3-1-3 平绣鸟纹，锁绣框边

Picture 3-1-3 Bird Pattern in Flat-stitch with Pin-embroidered Frame

图3-1-4 平绣花草纹，锁绣框边

Picture 3-1-4 Patterns of Flower and Grass in Flat-stitch with Pin-embroidered Frame

图3-1-5 平绣火焰纹，锁绣框边

Picture 3-1-5 Flame Patterns in Flat-stitch with Pin-embroidered Frame

图3-1-6　宽幅直筒裙

Picture 3-1-6　Wide Straight Skirt

图3-1-7　锁绣框边

Picture 3-1-7　Pin-embroidered Frame

» 名称：安顺岩蜡苗族无领对襟女套装

» 时代：约2000年

» 馆藏：W.1765

主要流行于贵州安顺、紫云、镇宁等地，以安顺岩蜡乡为代表。该套装由上衣、直筒裹裙、围腰、腰带、绑腿等组成。上衣以红色为主色调，多种色布、织锦拼接而成，肩背部拼接数纱绣的几何纹花块，下装为黑、白、蓝色等布条相间拼接的直筒裹裙，裙正中镶一条象征山川的数纱绣三角纹花块，穿着时合缝在前，形成“八”字，上系白色带穗素色长方形围腰，腰间束红色织锦带穗宽腰带。相传直筒裙上相间的黑、白条纹代表苗族迁徙途中翻越的山川和渡过的江河。此套装为当地妇女的盛装，便装时，上衣多为蓝、绿色，无挑花块装饰。

图3-2-1 安顺岩蜡苗族无领对襟女套装（正面）

Picture 3-2-1 Miao's Collarless Symmetric Woman Outfit at Yanla Village of Anshun Area (Front)

» Name: Miao's Collarless Symmetric Woman Outfit at Yanla Village of Anshun Area

» Time: About 2000

» Collection: W.1765

Represented by the style at Yanla Village of Anshun City, the outfit is popular in Anshun City, Ziyun County and Zhenning County in Guizhou Province. It is composed of jacket, straight skirt, apron, sash and leggings. Red as the main color, the jacket, full of yarn-counting stitched geometric patterns on shoulders and back, is made up of multi-colored cloth and brocade. The black, white and blue straight skirt is yarn-counting stitched with a triangular pattern symbolizing mountains and rivers, which is stitched in the front to form a shape of Chinese character "八（ba）". In addition, women wear rectangular white-cloth apron with a red brocaded sash. It is said that the black and white strips of the outfit represent the mountains and rivers the ancestors migrated. The outfit in the picture is woman festival costume. The jacket of daily dress is in blue and green without cross-stitched patterns.

图3-2-2 安顺岩蜡苗族无领对襟女套装（背面）

Picture 3-2-2 Miao's Collarless Symmetric Woman Outfit at Yanla Village of Anshun Area (Back)

图3-2-3 安顺岩蜡苗族女盛装（雷洪斌 摄）

Picture 3-2-3 Miao's Woman Festival Costume at Yanla Village of Anshun Area (Photographed by Lei Hongbin)

» 名称：安顺宁谷苗族对襟女套装
» 时代：约1970年
» 馆藏：W.1764

主要流行于贵州安顺宁谷、汪家山、猛邦周边，以安顺宁谷为代表。该套装由上衣、百褶裙、前围腰、后围腰、腰带、头帕、头带、绑腿等组成。上衣为家织青布对襟衣，门襟、双袖（分四段）上是平绣变形的鸟纹和几何纹，衣后摆绣两对凤凰和三个“喜”字，后衣领下的方块中间刺绣象征男女生殖器的葵纹，暗喻阴阳合一，是较为典型的生殖崇拜图案。下装为家织青布百褶裙，裙下摆绣葫芦花，饰粉白色条布。前围腰至膝间，上刺绣菱形纹花块，后围腰由六条刺绣宽花带拼接而成，一般多层表示富有。另有挑花头带、多层青布头帕和青布带挑花的绑腿。该服饰为嫁衣，女子佩戴有银项圈。

图3-3-1　安顺宁谷苗族对襟女套装（正面）

Picture 3-3-1　Miao’s Symmetric Woman Outfit at Ninggu Village of Anshun Area (Front)

图3-3-2 安顺宁谷苗族对襟女套装（背面）
Picture 3-3-2 Miao's Symmetric Woman Outfit at Ninggu Village of Anshun Area (Back)

图3-3-3　平绣葵纹

Picture 3-3-3　Flat-stitched Sunflower Patterns

图3-3-4　衣袖局部图

Picture 3-3-4　Sleeve (Section)

图3-3-5　平绣葫芦花纹

Picture 3-3-5　Flat-stitched Calabash Flower Patterns

» Name: Miao's Symmetric Woman Outfit at Ninggu Village of Anshun Area

» Time: About 1970

» Collection: W.1764

Represented by the style at Ninggu Village of Anshun City, the outfit is popular in villages of Ninggu, Wangjiashan, Mengbang of Anshun City in Guizhou Province. It is composed of jacket, pleated skirt, front and back aprons, sash, headscarf, headwear and leggings. Made of home-woven black cloth, the jacket is full of embroidered geometric patterns and transformed patterns of bird on lapel and sleeves, two pairs of phoenixes and three Chinese characters "喜（xi）" on jacket tail, and sunflower patterns representing genital organs on a square panel right below the back collar. The black-cloth pleated skirt is embroidered with calabash flower patterns decorated with pink white strips. The front apron is knee length and embroidered with diamond patterns while the back apron is made of six embroidered sashes stitched with the connotation of wealth. In addition, women wear cross-stitched headwear, multi-layer black-cloth headscarf and black-cloth cross-stitched leggings. The outfit in the picture is woman costume for wedding, ornamented with silver neck-rings.

图3-3-6　平绣双凤招喜

Picture 3-3-6　Flat-stitched "Two Phoenixes"

图3-3-7　平绣菱形纹花块

Picture 3-3-7　Flat-stitched Diamond Patterns

图3-3-8 安顺宁谷苗族女盛装（雷洪斌 摄）

Picture 3-3-8 Miao's Woman Festival Costume at Ninggu Village of Anshun Area (Photographed by Lei Hongbin)

» 名称：安顺长树角苗族右衽女套装
» 时代：约1950年
» 馆藏：W.1759

主要流行于贵州安顺、镇宁苗、布依杂居地，以安顺长树角为代表。该套装由上衣、百褶裙、围腰、腰带和头帕组成。上衣为褐色绸缎收腰夹衣，右衽门襟及双袖饰平绣折枝花、鸟纹，以栏杆饰之。衣下摆呈弧形，饰各色丝线包边的绸缎排列叠压，略显精致。由于当地几百年来传承着祖先崇拜龙的习俗，下装多为蜡染点状龙鳞纹的家织布百褶裙。围腰及腰带平绣花草纹。头帕中部为青布彩色织锦。盛装时，女子头上插一根银簪子。

图3-4-1　安顺长树角苗族右衽女套装

Picture 3-4-1　Miao’s Right-buttoned Woman Outfit at Changshujiao Village of Anshun Area

» Name: Miao's Right-buttoned Woman Outfit at Changshujiao Village of Anshun Area

» Time: About 1950

» Collection: W.1759

Represented by the style at Changjiaoshu Village of Anshun City, the outfit is popular in Anshun City and Miao and Puyi residential places of Zhenning County in Guizhou Province. It is composed of jacket, pleated skirt, apron, sash and headscarf. The jacket, full of embroidered patterns of flower and bird on right lapel and sleeves decorated with railing patterns, is a brown brocaded waist-tight lined coat. The coat tail, decorated with brocade hemmed by multi-colored silk lines, is in the shape of curve. The home-woven pleated skirt is dotted with batik dragon scale patterns due to the custom of worship for dragon. The apron and sash are embroidered with patterns of flower and grass. The middle of headscarf is black-cloth colored brocade. Women ornament themselves with silver hairpins at festivals.

图3-4-2　褐色绸缎收腰夹衣（背面）

Picture 3-4-2　Brown Satin Waist Jacket (Back)

图3-4-3　庆祝节日（雷洪斌　摄）

Picture 3-4-3　Celebrating Activities at Festivals (Photographed by Lei Hongbin)

» 名称：镇宁黄果树布依族对襟女套装
» 时代：约1960年
» 馆藏：W.1760

主要流行于贵州镇宁、关岭、安顺布依族聚居地，以镇宁黄果树、扁担山为代表。该套装由上衣、百褶裙、围腰、腰带、头帕等组成。上衣为家织青布对襟衣，门襟、双袖（分三段）、后背饰有漩涡纹、团花纹、铜鼓纹蜡染和织锦。下装为斑点纹、漩涡纹蜡染百褶长裙。裙上系青布织锦围腰，另有青布织锦头帕。镇宁黄果树布依族对襟衣上的蜡染漩涡纹、铜鼓纹代表着布依族人民对大自然、对铜鼓的崇拜，团花纹则是该土语区内各家族团结的象征。布依族崇拜自然、图腾和祖先，自然崇拜以敬奉灶神、山神、神石和神树等；图腾崇拜就体现在蜡染图案中漩涡纹和铜鼓纹等的使用；而祖先崇拜则希望他们的在天之灵能关心整个家族，保佑人丁兴旺，繁荣昌盛。因此，平时或节日宴请亲朋做好饭菜后，都要先供奉祖先。

图3-5-1　镇宁黄果树布依族对襟女套装（正面）
Picture 3-5-1　Puyi's Symmetric Woman Outfit at Huangguoshu Village of Zhenning (Front)

» Name: Puyi's Symmetric Woman Outfit at Huangguoshu Village of Zhenning

» Time: About 1960

» Collection: W.1760

Represented by the style at villages of Huangguoshu and Biandanshan in Anshun City, the outfit is popular in Puyi-populated areas of Counties of Zhenning and Guanling and Anshun City in Guizhou Province. It is composed of jacket, pleated skirt, apron, sash and headscarf. Made of home-woven black cloth and composed of symmetric coat, lapel and sleeves (three parts), the jacket is full of batik and brocaded patterns of eddy, flower and bronze drum on back. The long pleated skirt is decorated with batik patterns of spot and eddy. Women wear black-cloth brocaded aprons and headscarves. The patterns of eddy and bronze drum represent Puyi's worship for nature and bronze drum, while flower patterns for unity of different families in local. Puyi worship for nature (Kitchen God, Mountain God, divine stone and divine tree), totems (batik patterns of eddy and bronze drum) and ancestors (ancestors' spirits in heaven caring about the whole family and protecting the prosperity of the population). Therefore, after preparing meals for entertaining relatives and friends at ordinary times or at festivals, ancestors should be worshipped first.

图3-5-2 镇宁黄果树布依族对襟女套装（背面）

Picture 3-5-2 Puyi's Symmetric Woman Outfit at Huangguoshu Village of Zhenning (Back)

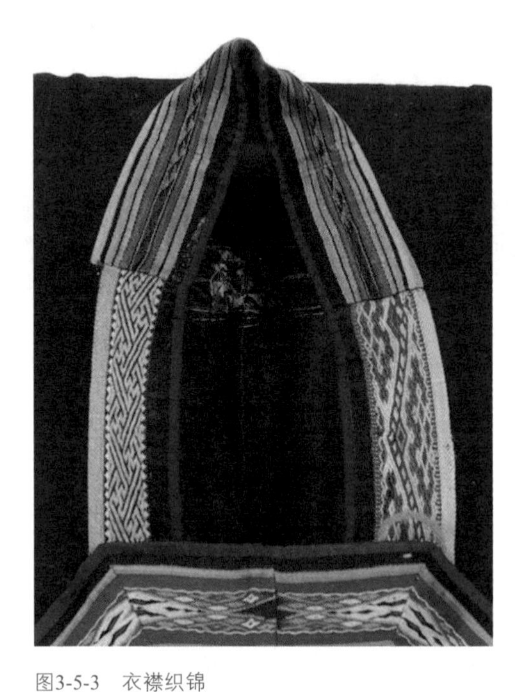

图3-5-3　衣襟织锦

Picture 3-5-3　Brocaded Lapel

图3-5-4　衣袖蜡染漩涡纹和菱形织锦

Picture 3-5-4　Batik Patterns and Brocaded Diamond Patterns on Sleeves

图3-5-5 平绣八角花几何纹

Picture 3-5-5 Flat Stitched Geometric Patterns of Octagon Flowers

图3-5-6 蜡染漩涡纹

Picture 3-5-6 Batik Eddy Patterns

图3-5-7 织锦几何纹

Picture 3-5-7 Brocaded Geometric Patterns

图3-5-8　安顺黄果树布依族女盛装（雷洪斌　摄）

Picture 3-5-8　Puyi's Woman Festival Costume at Huangguoshu Village of Anshun (Photographed by Lei Hongbin)

第四章
黔南地区少数民族服饰

Chapter 4 Ethnic Group's Costumes and Ornaments in South of Guizhou Province

» 名称：惠水蛮莫岩脚布依族女套装

» 时代：约1990年

» 馆藏：W.1772

1997 年 11 月，征集于惠水县太阳乡蛮莫村岩脚组杨彩维家。据杨彩维 90 多岁的婆婆陈罗氏讲述，这套衣服是她 18 岁出嫁陈家时母亲为她制作的嫁衣。

衣服染色是采集一种布依语叫“冬绿”的藤类新鲜叶茎，捶压榨挤，多次加温至沸浓缩成汁，待冷却后加入一定比例的滤好的青岗柴草木灰汁液，用小白布定好色调（可根据需要加入或减少控制色的深浅），将布料用温水浸湿待用，染缸用木棍搅动倒入烈酒激活，一个小时后将棉布用染架垂直放入染缸，浸染约一个小时，取出刷弃残留在布上的染渣，晾于阴干处，反复入染十余次，观察色彩满意后再浸入蓝靛染缸染一次即可。

布依族妇女着装：白色家织布做头帕，上衣为有领大襟衣，精美绣花围腰；下为长宽绣花裤，脚踏满花鞋，或为半爿绣花，或鞋尖绣小花。其间镶嵌点缀少许银饰，甚洁净淡雅，古朴端庄。而出阁少女大体与上相同，唯于头帕末端，喜镶艳丽之纹案于发顶簪间，半掩半露，甚透灵秀娇娆。每逢佳节盛会，更佩耳环、戒指、项圈、发坠、手镯等银饰于身。如今，木叶捶染技艺已绝于民间，此套衣物为仅存唯一实物。

» Name: Puyi's Woman Outfit at Yanjiao, Manmo Village of Huishui

» Time: About 1990

» Collection: W.1772

In Nov. 1997, this outfit was collected at Yang Caiwei's home, Yanjiao, Manmo Village, Taiyang Town of Huishui County. According to Mrs. Luo, Yang Caiwei's 90-year-old mother-in-law, the dress was a dowry made by her mother when she married the Chen family at the age of 18.

This outfit was dyed with the juice, which was made from the leaf of a kind of ivy named "donglv" in Puyi language after boiling and concentration and mixed with a proper portion of grass ash. During the dyeing process, the fabric was stirred with wine and "donglv" after being wetted completely for an hour. This step was repeated for no less than 10 times until the demand was met.

The outfit is composed of white-cloth headscarf, big lapel jacket with collar, embroidered apron, long and loose embroidered pants and shoes full of embroidered flower patterns, half embroidered or embroidered flower patterns on toe cap with a few silver ornaments decorated. Girls' outfits are a little bit different from married women's in a certain part: colorful patterns ornamented at the end of scarf and between hairpins, which look delicately beautiful. At festivals, women wear silver jewelries, including earrings, rings, neck-rings, hair drops and bracelets. Nowadays, the leaf dyeing technique has been lost and this outfit is the only physical costume left.

图4-1-1 惠水蛮莫岩脚布依族女套装（正面）

Picture 4-1-1 Puyi's Woman Outfit at Yanjiao, Manmo Village of Huishui (Front)

图4-1-2 惠水蛮莫岩脚布依族女套装（背面）

Picture 4-1-2 Puyi's Woman Outfit at Yanjiao, Manmo Village of Huishui (Back)

图4-1-3　布依族“染”色

Picture 4-1-3　Dyeing of Puyi

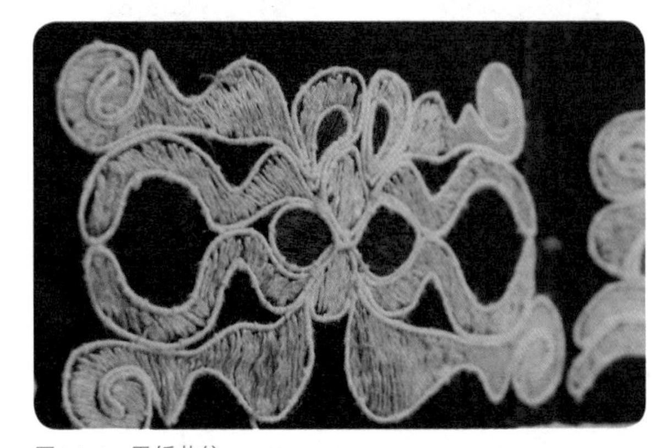

图4-1-4　平绣花纹

Picture 4-1-4　Flat-stitched Flower Patterns

图4-1-5　挑花鸟纹和花纹

Picture 4-1-5　Cross-stitched Patterns of Bird and Flower

图4-1-6 惠水岩脚女盛装（雷洪斌 摄）

Picture 4-1-6 Puyi's Woman Festival Costume at Yanjiao of Huishui (Photographed by Lei Hongbin)

» 名称：惠水鸭绒苗族交襟女套装
» 时代：约1950—1960年
» 馆藏：W.1771

主要流行于贵州惠水、长顺、平塘和龙里等地，以惠水鸭绒为代表。该套装由上衣、围腰和百褶裙组成。上衣为青布交襟衫，双袖饰刺绣几何纹，围腰胸部与腰带数纱绣折枝花，百褶长裙由多种色彩的家织布拼接而成，上镶嵌贴布绣卷草花、枫香染水车花和织锦几何纹样，花块错落有致，五彩斑斓，传说其中的三条横杠代表苗族祖先迁徙途中渡过的江河。该地区以枫香染工艺为特色，裙子中的贴布绣最为美观。传统的套装分为黑色和蓝色，蓝色多用于丧葬场合。

图4-2-1 惠水鸭绒苗族交襟女套装（正面）

Picture 4-2-1 Miao's Cross Lapel Woman Outfit at Yarong Village of Huishui (Front)

» Name: Miao's Cross Lapel Woman Outfit at Yarong Village of Huishui
» Time: About 1950s
» Collection: W.1771

Represented by the style at Yarong Village of Huishui County, the outfit is popular in Counties of Huishui, Changshun, Pingtang and Longli in Guizhou Province. It is composed of jacket, apron and pleated skirt. Made of black cloth, the jacket is full of embroidered geometric patterns on sleeves and flower patterns on apron and sash. The long pleated skirt is sewed with multi-colored home-woven cloths and applique-stitched with patterns of different flowers and brocaded geometric patterns. The patterns are well arranged and colorful. It is said that the three bars among the patterns stand for rivers Miao ancestors ever crossed in their migration. The skirt is famous for its maple juice dyeing and applique embroidery. The traditional outfits are black or blue, with the blue one usually for the funeral.

图4-2-2 惠水鸭绒苗族交襟女套装（背面）

Picture 4-2-2 Miao's Cross Lapel Woman Outfit at Yarong Village of Huishui (Back)

图4-2-3　平绣花纹

Picture 4-2-3　Flat Stitched Flower Patterns

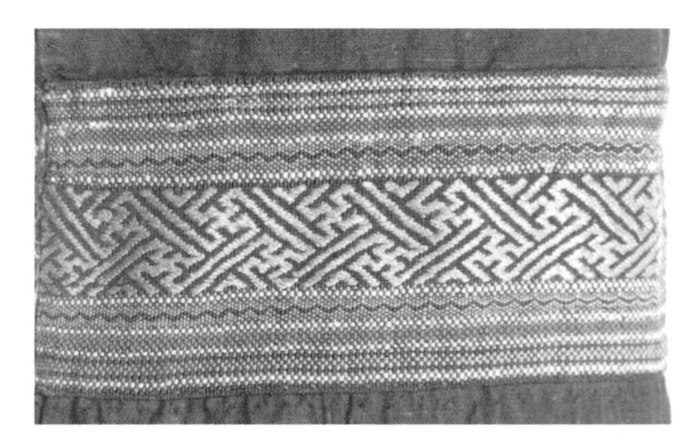

图4-2-4　织锦几何纹

Picture 4-2-4　Brocaded Geometric Patterns

图4-2-5　数纱绣折枝花

Picture 4-2-5　Flower Patterns in Yarn-counting Stitch

图4-2-6　枫香染水车花

Picture 4-2-6　Patterns of Waterwheel in Sweet Gum Dyeing

图4-2-7　贴布绣卷草花

Picture 4-2-7　Applique-stitched Flower Patterns

图4-2-8 惠水鸭绒苗族女盛装（雷洪斌 摄）

Picture 4-2-8 Miao's Woman Festival Costume at Yarong Village of Huishui (Photographed by Lei Hongbin)

» 名称：惠水摆金苗族无领对襟女套装

» 时代：约2000年

» 馆藏：W.1773

主要流行于贵州惠水摆金、甲浪、城关等地，以惠水摆金镇为代表。该套装由上衣、百褶裙、腰带、背牌、头帕、绑腿组成。上衣为红色绸缎无领对襟宽阔半袖衣，双袖轴线绣山茶花，间钉龙纹、铜鼓纹银泡装饰。衣领处饰有轴线绣山茶花压领，塑料珠穗子饰之。下装为青布百褶裙。腰带轴线绣山茶花，饰龙纹、铜鼓纹银泡，缀穗子和挂件。另应有青布绑腿、轴线绣背牌、穗子绣花鞋。盛装时，头挽发髻于顶，用约 2 米长、0.16 米宽的青布缠数圈，再戴绣花帽，帽檐四周缀无数银珠与玉珠，发髻上插两组由 4 条银片组成的银冠，银冠下两侧插有银花和银片，髻后插银梳，戴耳环。

图4-3-1　惠水摆金苗族无领对襟女套装（正面）

Picture 4-3-1　Miao's Symmetric Woman Outfit at Baijin Village of Huishui (Front)

» Name: Miao's Symmetric Woman Outfit at Baijin Village of Huishui

» Time: About 2000

» Collection: W.1773

Represented by the style at Baijin Village of Huishui County, the outfit is popular in villages of Baijin, Jialang and Chengguan of Huishui County in Guizhou Province. It is composed of jacket, pleated skirt, sash, back panel, headscarf and leggings. The red brocaded collarless symmetric jacket with loose sleeves is axis embroidered with flower patterns on sleeves and ornamented with patterns of dragon and bronze-drum like silver bubbles and axis embroidered flower patterns on collars and ornamented with tassels made of plastic beads. The black-cloth pleated skirt is matched with a sash with axis embroidered patterns of camellia, dragon, bronze-drum like silver bubbles, tassels and pendants. Women wear black-cloth leggings, back panel in axis embroidery and embroidered shoes with tassels. On festival occasions, women wear topknots with embroidered caps ornamented with countless silver beads and jade beads after wrapping with 2-meter long, 0.16-meter wide black cloths. In addition, women decorate themselves with silver crowns composed of four pieces of silver strips, with silver flowers and silver strips at two bottom sides and silver combs and earrings.

图4-3-2　塑料珠穗帽子

Picture 4-3-2　Cap with Tassels Made of Plastic Beads

图4-3-3　轴线绣山茶花

Picture 4-3-3　Axis Embroidered Camellia Patterns

图4-3-4　银锁

Picture 4-3-4　Silver Lock

图4-3-5 塑料珠穗子

Picture 4-3-5 Tassel Made of Plastic Beads

图4-3-6 惠水摆金女盛装（雷洪斌 摄）

Picture 4-3-6 Miao's Woman Festival Costume at Baijin Village of Huishui (Photographed by Lei Hongbin)

» 名称：贵定新场苗族对襟女套装

» 时代：约1990年

» 馆藏：W.242

主要流行于贵州贵定、福泉、开阳等地，以贵定新场为代表。该套装由上衣、百褶裙、围腰、背牌、头帕、腰带、绑腿组成。上衣为青布翻领对襟衣，两衣襟、双袖、肩部对称饰有平绣、贴布绣、锁绣、挑花混合组成的黄豆花、栀子花、云彩、方块花、菱形花等图案。下装为蜡染百褶裙，中间部分以红、褐色绸缎点缀。围腰以各色布、栏杆条镶嵌成半回纹，中间锁绣黄豆花。背牌挑绣扣子花、四叶花、织锦几何纹等，并缀23颗海贝，海贝与刺绣相互映衬，极为精致。另应有蜡染头帕、青布绑腿等。

图4-4-1　贵定新场苗族对襟女套装（正面）

Picture 4-4-1　Miao’s Symmetric Woman Outfit at Xinchang Village of Guiding（Front)

» Name: Miao's Symmetric Woman Outfit at Xinchang Village of Guiding
» Time: About 1990
» Collection: W.242

Represented by the style at Xinchang Village of Guiding County, the outfit is popular in counties of Guiding, Fuquan and Kaiyang in Guizhou Province. It is composed of jacket, pleated skirt, apron, back panel, headscarf, sash and leggings. The symmetric black-cloth jacket with turn-down collars is embroidered with patterns of different flowers and geometric patterns on lapels, sleeves and shoulders. The batik pleated skirt is decorated with red and brown brocade. With embroidered flower patterns in the middle, the apron is decorated with multi-colored cloth strips and half "回" (hui) pattern-like railing embroideries. Decorated with 23 seashells, back ornamentation is cross-stitched with button flowers, four-leaf flowers and brocaded geometric patterns. Women wear batik turbans and black-cloth leggings as well.

图4-4-2 贵定新场苗族对襟女套装（背面）

Picture 4-4-2 Miao's Symmetric Woman Outfit at Xinchang Village of Guiding (Back)

图4-4-3　衣襟局部图

Picture 4-4-3　Lapel (Section)

图4-4-4　衣袖局部图

Picture 4-4-4　Sleeve (Section)

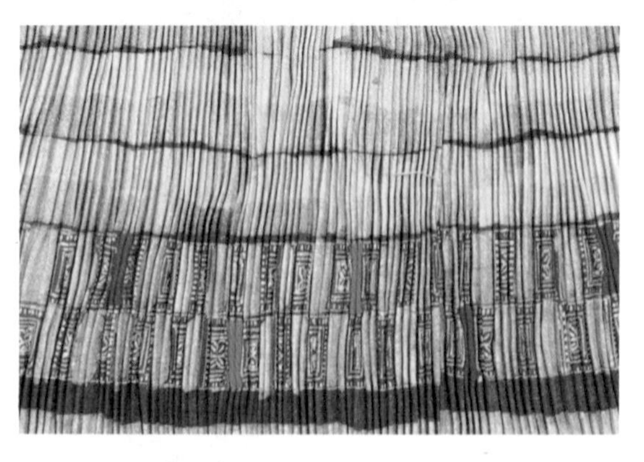

图4-4-5　百褶裙局部图

Picture 4-4-5　Pleated Skirt (Section)

图4-4-6　平绣、锁绣、挑花几何纹和花纹

Picture 4-4-6　Geometric Patterns and Flower Patterns in Flat, Pin and Cross Stitch

图4-4-7　平绣、锁绣花纹

Picture 4-4-7　Flower Patterns in Flat and Pin Stitch

图4-4-8　挑花、锁绣花草纹

Picture 4-4-8　Patterns of Flower and Grass in Flat and Pin Stitch

图4-4-9 贵定苗族女盛装

Picture 4-4-9 Miao's Woman Festival Costume in Guiding

» 名称：平塘油岜苗族无领贯首女套装

» 时代：约1950年

» 馆藏：W.250

主要流行于贵州平塘油岜、鼠场、金桥及广西南丹等地，以贵州平塘油岜乡为代表。该套装由上衣、百褶裙、腰带、绑腿、手帕组成。上衣为青布无领贯首衣，衣襟前短后长，双袖及前后衣襟上挑花几何纹、方块纹和十字交叉纹，后衣襟缝接红色织锦半回纹衣片。下装为家织青布百褶裙，裙摆下缝接红色织锦，并在上方挑花代表支系的人字花。另有双色挑花腰带、白布绑腿和流苏花手帕。相传平塘油岜苗族首领蚩尤涿鹿之战败后，后裔经过四次大的迁徙到达贵州，为纪念祖先，妇女们在自己的衣服上刺绣了代表蚩尤印章象征符号的方块纹，还有迁徙途中狩猎用的弓箭和祖先曾经居住过的地方。

图4-5-1　平塘油岜苗族无领贯首女套装（正面）

Picture 4-5-1　Miao's Collarless Woman Pullover at Youba Town of Pingtang (Front)

» Name: Miao's Collarless Woman Pullover at Youba Town of Pingtang

» Time: About 1950

» Collection: W.250

Represented by the style at Youba Town of Pingtang County, the outfit is popular in towns of Youba, Shuchang and Jinqiao of Pingtang County in Guizhou Province and in Nandan City of Guangxi. It is composed of jacket, pleated skirt, sash, leggings and handkerchief. With short front and long back lapels, the collarless black-cloth pullover is cross-stitched with geometric patterns , square patterns and cross patterns on sleeves and sewed with red brocaded half "回" (hui) patterns. The home-woven black-cloth pleated skirt is stitched with red brocade and cross-stitched with "人"(ren) patterns which represent its branches. Women wear two-colored cross-stitched sash, white leggings and handkerchief with tassels and flower patterns. According to legend, the ancestor of the branch of Miao people inhabiting in the area of Youba of Pingtang was Warrior Chiyou, a leader of Jiuli tribe. After his defeat in the battle of zhuolu, the descendants migrated to Guizhou Province. The square patterns embroidered on the jacket are in memory of their ancestor, the bows and arrows used in their hunting and the places they've lived.

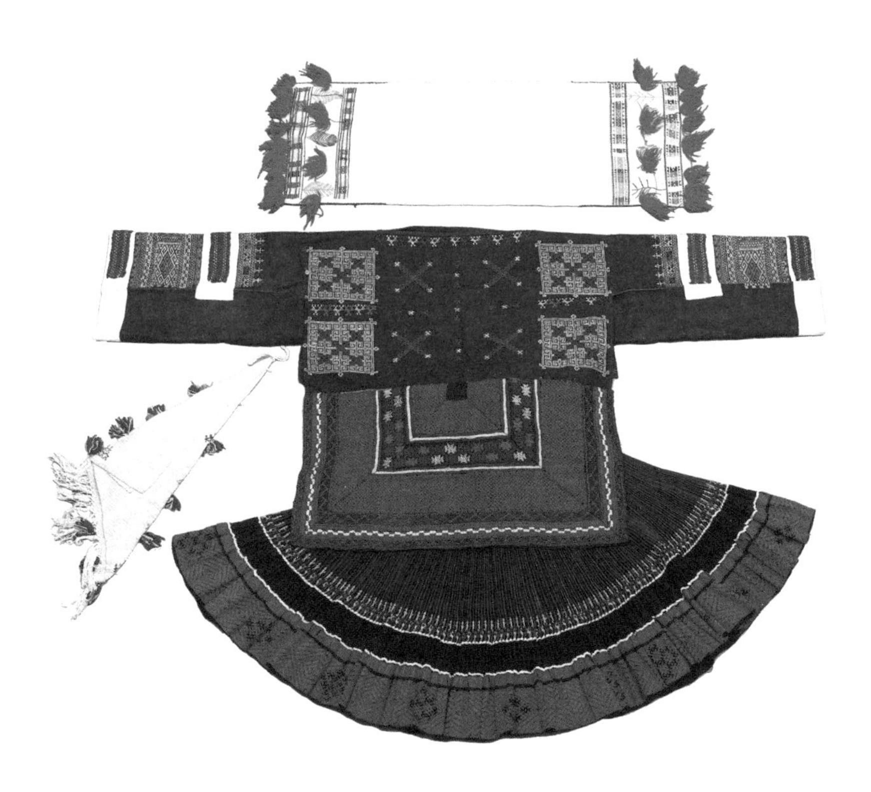

图4-5-2 平塘油岜苗族无领贯首女套装（背面）

Picture 4-5-2 Miao's Collarless Woman Pullover at Youba Town of Pingtang (Back)

图4-5-3　挑花几何纹

Picture 4-5-3　Cross-stitched Geometric Patterns

图4-5-4　挑花方块纹、十字交叉纹

Picture 4-5-4　Cross-stitched Patterns of Square and Cross

图4-5-5　挑花十字交叉纹

Picture 4-5-5　Cross-stitched Cross Patterns

图4-5-6　家织青布百褶裙局部图

Picture 4-5-6　Pleated Skirt Made of Home-woven Black Cloth (Section)

图4-5-7 平塘苗族女盛装（雷洪斌 摄）

Picture 4-5-7 Miao's Woman Festival Costume at Pingtang County (Photographed by Lei Hongbin)

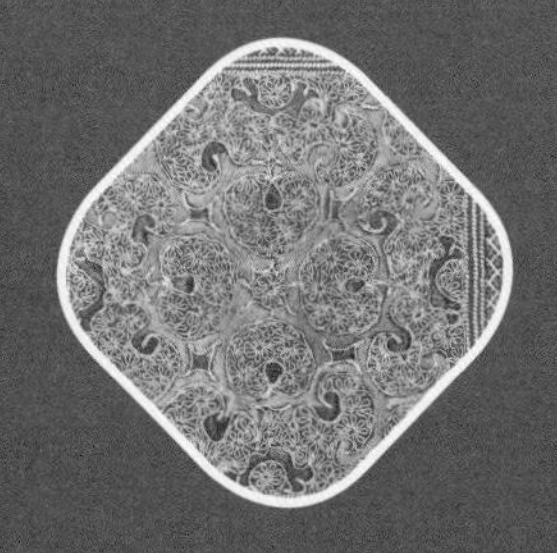

第五章

毕节地区少数民族服饰

Chapter 5 Ethnic Group's Costumes and Ornaments in Bijie Area

» 名称：金沙苗族背褡式对襟女套装
» 时代：约1970年
» 馆藏：W.1730

主要流行于金沙及四川叙永、古蔺等地，以四川叙永麻城乡为代表。该套装由上衣、百褶裙、胸牌、云肩、腰带、头帕、绑腿等组成。上衣为青布背褡式对襟夹衣，前后下摆以黄色为主调挑花蜘蛛纹，后领处缝接长方形挑花背褡，上套拼布八角星纹的云肩。相传挑花背褡象征祖先居住的城郭。下装为织锦、贴布绣、蜡染相间的百褶长裙。腰间系带穗的彩色织锦腰带，头缠黑白织锦头帕，打白布绑腿。另外还有挑花八角星纹象征蚩尤九黎符号的方形胸牌。

图5-1-1 金沙苗族背褡式对襟女套装
Picture 5-1-1 Miao's Waistcoat-Style Symmetric Woman Outfit at Jinsha County

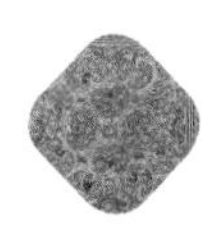

» Name: Miao's Waistcoat-Style Symmetric Woman Outfit at Jinsha County

» Time: About 1970

» Collection: W.1730

Represented by the style at Macheng Town, Xuyong County of Sichuan Province, the outfit is popular in Jinsha County in Guizhou, Xuyong County and Gulin County in Sichuan Province. It is composed of jacket, pleated skirt, breast ornamentation, shoulder adornment, sash, headscarf and leggings. The jacket is a black-cloth waistcoat-style symmetric coat with yellow cross-stitched spider patterns on lower hem, sewed with a rectangular cross-stitched waistcoat at back collar and covered with a shoulder adornment with octagonal star patterns. According to legend, the cross-stitched waistcoat symbolizes the city walls the ancestors lived. The brocaded, applique-stitched and batik pleated skirt is matched with colorful brocaded sash with tassels. Women wear white and black brocaded headscarves, white leggings, and square breast ornamentation with cross-stitched octagonal star patterns which represent Miao's ancestor "Chiyou" (warrior).

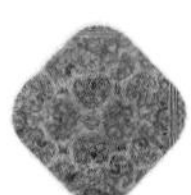

图5-1-2 拼布八芒花云肩

Picture 5-1-2 Shoulder Adornment with Patched Flower Patterns

图5-1-3　挑花蜘蛛纹

Picture 5-1-3　Cross-stitched Spider Patterns

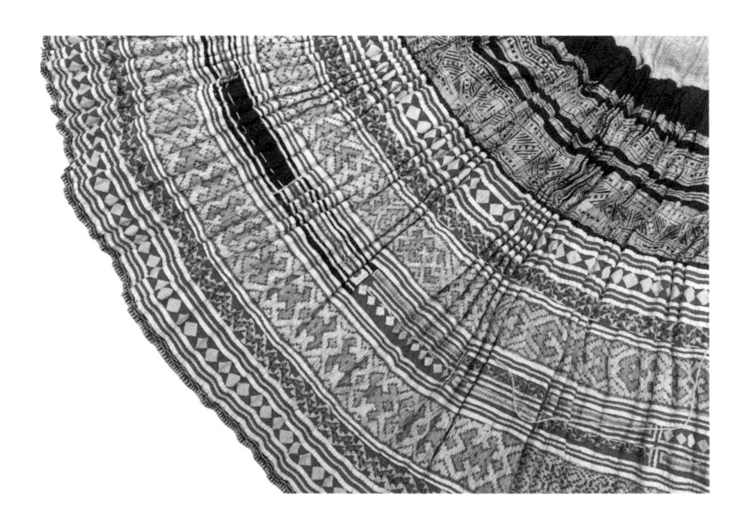

图5-1-4　织锦、贴布绣、蜡染相间百褶裙局部图

Picture 5-1-4　Brocaded, Applique-stitched and Batik Pleated Skirt (Section)

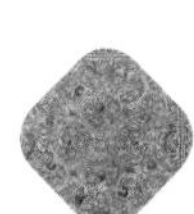

» 名称：威宁苗族披肩式无领对襟男套装

» 时代：约1960年

» 馆藏：W.1769

主要流行于贵州威宁、赫章、大方、毕节、黔西等地，以贵州威宁为代表。该套装由上衣、披肩、裙子、腰带、绑腿等组成。上衣为家织白布无领对襟衣，外披红、黑色羊毛线编织的披肩，披肩后缝接处缀有象征祖先居住过的城池背褡，背褡下饰有彩色绒球穗。下装为白色家织麻布蜡染裙，裙边蜡染饰之，裙身饰有八组红、绿、黄色横杠。另配有麻布腰带和绑腿。该套装披肩上有三角形、斜“井”字图案，分别代表祖先迁徙途中翻越的山川和祖先耕种过的田地；以几组扁棱形格连续纹为主干框架，图案上以棱形象征箭头，方格挑花象征山川、田园树木、星星、鱼等。裙身上的红色三条横杠代表祖先迁徙途中渡过的黄河、长江，以及走过的平原，该套装被称为田园山川服。

图5-2-1　威宁苗族披肩式无领对襟男套装（正面）

Picture 5-2-1　Miao's Collarless Symmetric Man Outfit with a Shawl at Weining County (Front)

图5-2-2　威宁苗族披肩式无领对襟男套装（背面）

Picture 5-2-2　Miao's Collarless Symmetric Man Outfit with a Shawl at Weining County (Back)

» Name: Miao's Collarless Symmetric Man Outfit with a Shawl at Weining County

» Time: About 1960

» Collection: W.1769

Represented by the style at Weining County, the outfit is popular in counties of Weining, Hezhang, Dafang, Qianxi and Bijie City in Guizhou Province. It is composed of jacket, shawl, skirt, sash and leggings. The jacket is a home-woven white-cloth symmetric coat with a shawl woven in red and black wool. Sewed with a back adornment stitched with towns and walls at the back, the shawl is decorated with colored fuzzy-ball tassels. Made of home-woven white cloth, the skirt is decorated with batik at the hem, eight groups of red, green and yellow horizontal bars. Women wear linen sash and leggings. The patterns of triangles and italic "井" (jing) on shawl respectively represent the mountains and fields ancestors climbed over and plowed. Several groups of flat prismatic lattice continuous patterns as frame, prismatic patterns symbolize arrows and square patterns symbolize mountains, rivers, fields, gardens, trees, stars and fish. The three horizontal red bars on skirt represent Yellow River, The Yangtze and plains ancestors passed in migration. The outfit is also called "Pastoral Landscape Costume".

图5-2-3 背褡

Picture 5-2-3 Shawl

图5-2-4 衣袖局部

Picture 5-2-4 Sleeve (Section)

图5-2-5 平绣斜“井”纹

Picture 5-2-5 Stitched Italic “井” Patterns

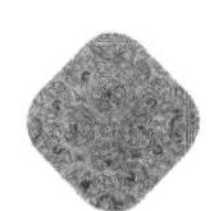

图5-2-6 威宁苗族男盛装

Picture 5-2-6 Miao’s Man Festival Costume at Weining County

» 名称：威宁苗族披肩式无领对襟女套装
» 时代：约1960年
» 馆藏：W.1770

主要流行于贵州威宁、赫章、大方、毕节、黔西等地，以贵州威宁为代表。该套装由上衣、披肩、裙子、腰带、绑腿等组成。上衣为家织白布无领对襟衣，外披红、黑色羊毛线编织的披肩，披肩后缝接处缀有象征祖先居住过的城池背褡，背褡下饰有彩色绒球穗。下装为白色家织麻布蜡染裙，裙边蜡染饰之，裙身饰有八组红、绿、黄色横杠。另配有麻布腰带和绑腿。该套装披肩上有三角形、斜“井”字图案，分别代表祖先迁徙途中翻越的山川和祖先耕种过的田地；裙身上的红色三条横杠代表祖先迁徙途中渡过的黄河、长江，以及走过的平原，该套装被称为田园山川服。

图5-3-1 威宁苗族披肩式无领对襟女套装

Picture 5-3-1 Miao's Collar-less Symmetric Woman Outfit with a Shawl at Weining County

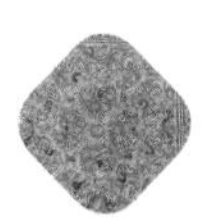

» Name: Miao's Collarless Symmetric Woman Outfit with a Shawl at Weining County

» Time: About 1960

» Collection: W.1770

Represented by the style at Weining County, the outfit is popular in counties of Weining, Hezhang, Dafang, Qianxi and Bijie City in Guizhou Province. It is composed of jacket, shawl, skirt, sash and leggings. The jacket is a home-woven white-cloth symmetric coat with a shawl woven in red and black wool. Sewed with a back adornment stitched with towns and walls at the back, the shawl is decorated with colored fuzzy-ball tassels. Made of home-woven whitecloth, the skirt is decorated with batik at the hem, eight groups of red, green and yellow horizontal bars. Women wear linen sash and leggings. The patterns of triangles and italic "井" (jing) on shawl respectively represent the mountains and fields ancestors climbed over and plowed. The three horizontal red bars on skirt represent Yellow River, The Yangtze and plains ancestors passed in migration. The outfit is also called "Pastoral Landscape Costume".

图5-3-2 威宁苗族披肩式无领对襟上衣局部图

Picture 5-3-2 Miao's Collarless Symmetric Woman Jacket with a Shawl at Weining County (Section)

图5-3-3 威宁苗族女盛装（右）

Picture 5-3-3 Miao's Woman Festival Costume at Weining County (Right)

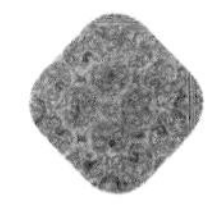

» 名称：纳雍百兴苗族对襟长尾女套装
» 时代：约1980年
» 馆藏：W1768

主要流行于贵州纳雍百兴乡、六枝特区等地，以百兴乡为代表。该套装由上衣、百褶裙、围腰、腰带、绑腿、头帕等组成。上衣为彩色蜡染对襟长尾衫，双袖、前襟彩色蜡染代表田园和犁耙的几何纹，长尾后背刺绣花椒花，中间等距离相间数纱绣几何纹。下装为家织布百褶裙，裙上段饰彩色蜡染，下段以红色布条饰之，代表祖先迁徙途中渡过的江河。腰间系青布百褶围腰，织锦镶嵌四周。另应有青布头帕、白布绑腿和腰带。盛装时，女子头卡红色雁型大木梳，四周缠一公斤重的假发，以毛麻挽髻，挽假发时须将梳角露出，脚穿绣花鞋，佩戴银手镯和耳环。

图5-4-1　纳雍百兴苗族对襟长尾女套装（正面）

Picture 5-4-1　Miao's Symmetric Woman Outfit with a Long Tail at Nayong County (Front)

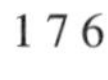

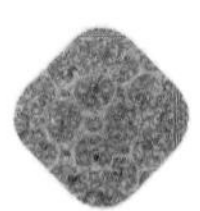

- Name: Miao's Symmetric Woman Outfit with a Long Tail at Nayong County
- Time: About 1980
- Collection: W.1768

Represented by the style at Baixing Town, the outfit is popular in Baixing Town of Nayong County and Liuzhi Special Administrative Region in Guizhou Province. It is composed of jacket, pleated skirt, apron, sash, leggings and headscarf. With colored batik geometric patterns on sleeves and lapel, yarn-counting stitched pepper flower patterns on tail and geometric patterns in the middle, the jacket is a colored batik symmetric coat with long tail. The home-woven pleated skirt is decorated with colored batik at the upper and red cloth strips at the bottom which represent rivers ancestors crossed during migration. Women wear black-cloth pleated apron adorned with brocade around, black-cloth headscarf, white leggings and sash. Women keep their hair and wig around big saddle-shaped red wooden combs, wear embroidered shoes and silver bracelets and earrings at festivals.

图5-4-2　纳雍百兴苗族对襟长尾女套装（背面）

Picture 5-4-2　Miao's Symmetric Woman Outfit with a Long Tail at Nayong County (Back)

图5-4-3　纳雍苗族女盛装（陈月巧　摄）

Picture 5-4-3　Miao's Woman Festival Costume at Nayong County (Photographed by Chen Yueqiao)

» 名称：大方竹园乡显母寨苗族背褡式对襟女套装

» 时代：约2000年

» 馆藏：W.1780

主要流行于贵州大方、织金、纳雍、黔西、毕节等地，以大方竹园乡显母寨为代表。该套装由上衣、百褶裙、披肩、前后围腰、腰带、绑腿等组成。上衣为拼布、挑花背褡式对襟长衫，双袖及肩部彩色拼布、挑花几何纹为主，前襟白布于胸前交叉，拴系于后腰。下装为蜡染饰彩条百褶裙，蜡染图案中包含山川、河流、井水等。盛装时，女子上身披带彩色线球的绣花披肩，披肩带交叉于胸前，系于后腰。腰间系多层素色前后围腰和白色宽腰带，打青布绑腿，头挽高大的三角形发髻，以巨大的木梳支撑。这是已婚女子装，百褶裙底边增加 7 寸的青布作为标记。

图5-5-1　大方竹园乡显母寨苗族背褡式对襟女套装

Picture 5-5-1　Miao’s Waistcoat-Style Symmetric Woman Outfit at Xianmu Village, Zhuyuan of Dafang County

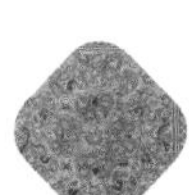

» Name: Miao's Waistcoat-Style Symmetric Woman Outfit at Xianmu Village, Zhuyuan of Dafang County

» Time: About 2000

» Collection: W.1780

Represented by the style at Xianmu Village, Zhuyuan of Dafang County, the outfit is popular in counties of Dafang, Zhijin, Nayong Qianxi and Bijie City in Guizhou Province. It is composed of jacket, pleated skirt, shawl, front and back aprons, sash and leggings. Cloths sewed, the jacket is a cross-stitched waistcoat-style symmetric long costume, decorated with colored cloths and cross-stitched geometric patterns on sleeves and shoulders. White lapels are crossed in front of chest and tied at back of waist. The pleated skirt in colored strips has different batik patterns including mountains, rivers and water in well. Women wear embroidered shawls with colored wool balls crossed in front of breast and tied at the back of waist. Hair in huge triangle-shape buns and adorned by wooden combs, women also wear multi-layer front and back aprons in single color, white wide sash and black-cloth leggings. This outfit is worn by married women with 7-inch black cloths added at the hem of skirt.

图5-5-2 衣袖贴布绣几何纹

Picture 5-5-2 Applique-stitched Geometric Patterns on Sleeves

图5-5-3 衣襟处平绣、蜡染、贴布绣等多种工艺

Picture 5-5-3 Flat Embroidery, Batik and Applique Embroidery on the Lapel

图5-5-4　大方竹园女盛装(雷洪斌　摄)

Picture 5-5-4　Miao’s Woman Festival Costume at Zhuyuan of Dafang County (Photographed by Lei Hongbin)

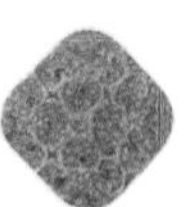

» 名称：大方八堡苗族大袖半臂对襟长尾男盛装
» 时代：约1970年
» 馆藏：W.1778

主要流行于贵州大方、毕节等地，以大方八堡乡为代表。该男装由上衣、阔腿裤和挑花腰带组成。相传该支苗族祖先以喜鹊为本支系的图腾加以崇拜，男女服饰仿照喜鹊黑白羽毛颜色制作，故称“鸦鹊苗”。上装除了在颜色上仿照喜鹊的黑色羽毛颜色外，在款式的裁剪上也模仿了喜鹊的长尾巴。下装为家织粗布阔腿裤。腰带较宽，各色方块挑花饰之，工艺精湛，色彩绚丽。该服饰与中国唐代宫廷中的半臂服颇为相似，是研究中国古代服饰款式的宝贵实物资料。

图5-6-1 大方八堡苗族大袖半臂对襟长尾男盛装
Picture 5-6-1 Miao's Symmetric Man Outfit with Loose and Half Sleeves and a Long Tail at Bapu of Dafang County

» Name: Miao's Symmetric Man Outfit with Loose and Half Sleeves and a Long Tail at Bapu of Dafang County

» Time: About 1970

» Collection: W.1778

Represented by the style at Bapu Town, the outfit is popular in Dafang County and Bijie City in Guizhou Province. It is composed of jacket, loose pants and cross-stitched sash. It is said that this branch of Miao ancestors' worship magpie as their totem, and the color design and style tailoring of the jacket is made to mimic white and black feathers and long tail of magpie. Therefore, this branch is known as "Yaque Miao" (Magpie Miao). The loose pants are made of home-woven cloths. The sash is wide and decorated with exquisite colorful cross-stitch. This outfit is quite similar to the half-sleeve court dress in Tang Dynasty. It is a valuable physical material for studying the style of ancient Chinese costumes.

图5-6-2 挑花腰带

Picture 5-6-2 Cross-stitched Belt

图5-6-3 大方八堡男女嫁装(陈月巧 摄)

Picture 5-6-3 Wedding Dresses for Men and Women at Bapu of Dafang County (Photographed by Chen Yueqiao)

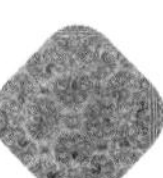

» 名称：大方八堡苗族大袖半臂对襟女盛装
» 时代：约1970年
» 馆藏：W.1777

主要流行于贵州大方、毕节等地，以大方八堡为代表。该套装由上衣、百褶裙、围腰、背褡及白布绑腿组成。相传该支苗族祖先以喜鹊为本支系的图腾加以崇拜，男女服饰仿照喜鹊黑白羽毛颜色制作，故称“鸦鹊苗”。女子上衣为黑白家织粗棉布制作的半臂对襟服，衣褂由正身、大袖、小袖、领四部分构成，正身两幅合缝，前襟分开，两边各接一副大袖，袖笼齐肩，后肩缝一副方形翻领。下装为青布百褶裙，裙下摆可以是红、白主色调为主的挑花及缎子方块作装饰，色调绚丽多彩。裙下摆也可以用蜡染布拼接而成，素雅端庄。裙子的皱褶约有 1200 个褶，三截镶成。

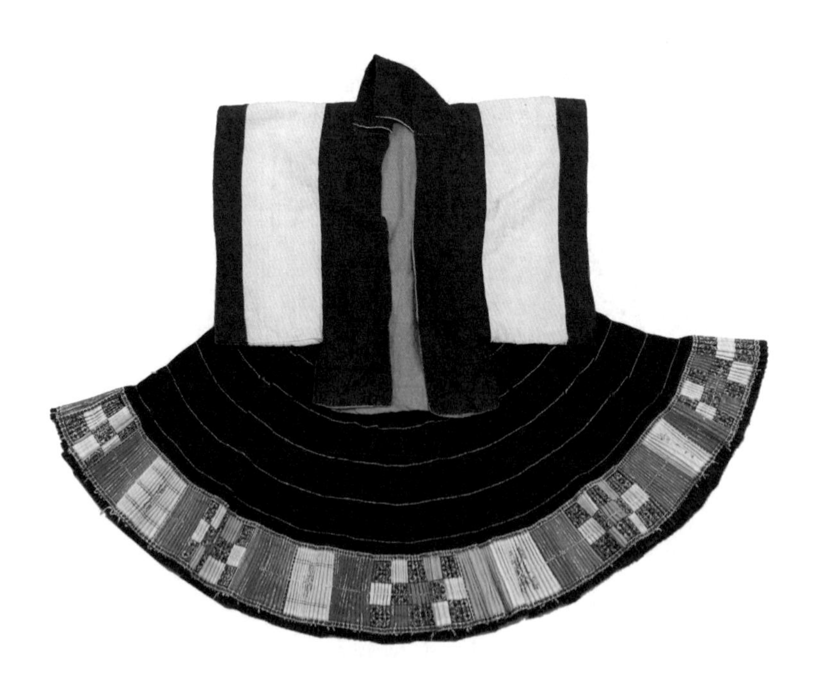

图5-7-1　大方八堡苗族大袖半臂对襟女盛装

Picture 5-7-1　Miao's Symmetric Woman Festival Costume with Loose and Half Sleeves at Bapu of Dafang County

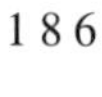

» Name: Miao's Symmetric Women Outfit with Loose and Half Sleeves at Bapu of Dafang County

» Time: About 1970

» Collection: W.1777

Represented by the style at Bapu Town, the outfit is popular in Dafang County and Bijie City in Guizhou Province. It is composed of jacket, pleated skirt, apron, waistcoat and white leggings. It is said that this branch of Miao ancestors' worship magpie as their totem, and the jacket is made to mimic white and black feathers of magpie. Therefore, this branch is known as "Yaque Miao" (Magpie Miao). The home-woven black and white jacket is a symmetric coat with half sleeves and composed of four parts including body, big sleeves, small sleeves and collar. With the lapel apart, the body part is sewed with two halves, stitched with a pair of loose sleeves at shoulder length and a square turndown collar at the back shoulder. The black-cloth pleated skirt is either decorated with colorful cross-stitched and brocaded square patterns in red and white as main color or sewed with simple but elegant batik cloth. The skirt is sewed with three sections and has about 1200 pleats.

图5-7-2　宽幅半袖上衣

Picture 5-7-2　Jacket with Loose and Half sleeves

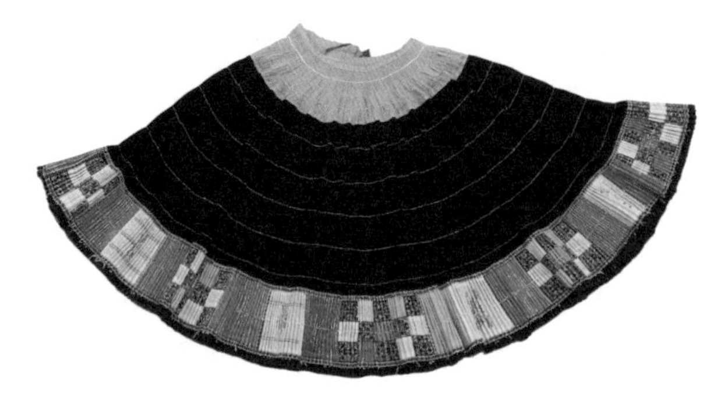

图5-7-3　三截拼接短裙

Picture 5-7-3　Mini Pleated Skirt Sewed with Three Sections

图5-7-4　裙身拼布绣

Picture 5-7-4　Patchwork Embroidery on Skirt

图5-7-5　青布百褶

Picture 5-7-5　Black-cloth Pleats on Skirt

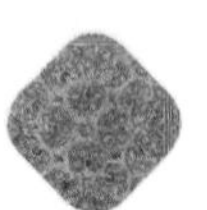

» 名称：织金箐脚苗族对襟女套装
» 时代：约1990年
» 馆藏：W.1781

主要流行于贵州织金、黔西等地，以织金箐脚乡为代表。该套装由上衣、直筒裹裙、围腰、绑腿等组成。上衣为黑色绒布对襟衣，双袖、门襟饰蜡染、贴布绣花草纹，后衣襟上下左右均饰醒目的贴布绣，上衣的下摆左右双层向上翘起，为穿衣者起到收腰的视觉效果，双袖蜡染彩色花草纹，衣襟及后背轴线绣云纹。下装为2米宽的百褶直筒裹裙，裙上以各色绸缎、贴布绣、彩色蜡染及轴线绣拼接装饰。腰间系黑底白条纹、四周以藤蔓纹、卷云纹蜡染装饰的长方形围腰。

图5-8-1　织金箐脚苗族对襟女套装（正面）
Picture 5-8-1　Miao's Symmetric Woman Outfit at Qingjiao of Zhijin County (Front)

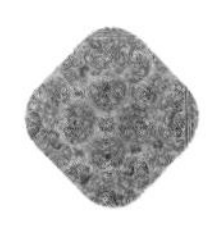

» Name: Miao's Symmetric Woman Outfit at Qingjiao of Zhijin County

» Time: About 1990

» Collection: W.1781

Represented by the style at Qingjiao Town, the outfit is popular in Zhijin County and Qianxi County in Guizhou Province. It is composed of jacket, straight skirt, apron and leggings. The black flannel symmetric jacket is decorated with batik and applique-stitched patterns on sleeves and lapel, and outstanding applique-stitched patterns on back lapel. The hem of the jacket is tilted up from the left and right sides to give the wearer a view of waist retraction. There are batik colored patterns of flower and grass on sleeves and axis embroidered cloud patterns on lapels and back. The two-meter wide straight pleated skirt is adorned with patterns stitched with colored brocade, applique, colored batik and axis embroidery. Women wear a rectangular black and white apron decorated with batik vines and cirrus patterns around.

图5-8-2　织金箐脚苗族对襟女套装（背面）

Picture 5-8-2　Miao's Symmetric Woman Outfit at Qingjiao of Zhijin County (Back)

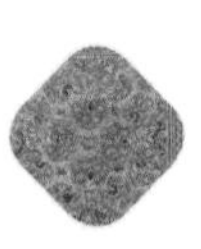

图5-8-3 织金箐脚女盛装(雷洪斌 摄)
Picture 5-8-3 Miao's Woman Festival Costume at Qingjiao of Zhijin County (Photographed by Lei Hongbin)

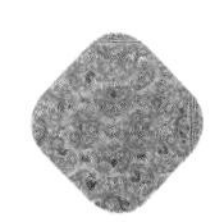

» 名称：织金珠场乡苗族轴线绣右衽女套装
» 时代：约2000年
» 馆藏：W.1761

主要流行于贵州织金、纳雍、大方、黔西等地，以织金珠场乡为代表。该套装由上衣、百褶直筒裹裙、围腰等组成。上衣为轴线绣右衽衣，双袖轴线绣火镰花和花椒花等，双袖另以黑、白、红、蓝色布饰之。下装为青布百褶直筒裹裙，裙上镶轴线绣火镰花、花椒花绣片和蜡染布条。裙外系轴线绣火镰花、花椒花等纹饰方围腰。整套服饰以醒目的黄色轴线绣为主，精细而简约，所谓的轴线绣是将丝线破成细股缠在一根棉线或两三根马尾上，再用缠好的梗线在布上绕成各种图案，以针线固定。便装时少用刺绣，纹饰不变，但是以蜡染工艺实现，其工艺称得上最精细的蜡染，线条流畅精细匀称。

图5-9-1 织金珠场乡苗族轴线绣右衽女套装（正面）

Picture 5-9-1 Miao's Right-buttoned Woman Outfit at Zhuchang of Zhijin County (Front)

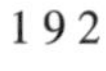

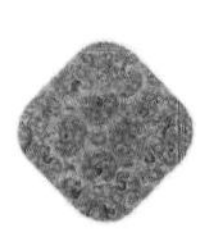

» Name: Miao's Right-buttoned Woman Outfit at Zhuchang of Zhijin County

» Time: About 2000

» Collection: W.1761

Represented by the style at Zhuchang Town, the outfit is popular in counties of Zhijin, Nayong, Dafang and Qianxi in Guizhou Province. It is composed of jacket, straight skirt and apron. The left-buttoned jacket is axis embroidered with patterns of fire sickle flower and pepper flower on sleeves which are decorated black, white, red and blue cloths. The black-cloth straight pleated skirt is also axis embroidered with fire sickle flower patterns, pepper flower patterns and batik cloth strips. Women wear square aprons axis stitched with patterns of fire sickle flower and pepper flower. The whole costume is featured in yellow axis embroidery which is to wind split sewing silk thread on a piece of cotton thread or on 2 to 3 horse tails, then wind them into various patterns and stitch them on the cloth. The outfit was made in the start of this century. The daily dress seldom uses embroidery, but keeps the patterns made by batik, one of the finest technique of Miao.

图5-9-2　织金珠场乡苗族轴线绣右衽女套装（背面）

Picture 5-9-2　Miao's Right-buttoned Woman Outfit at Zhuchang of Zhijin County (Back)

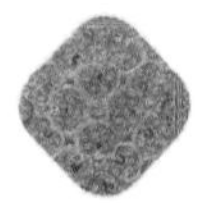

图5-9-3 织金珠场女盛装(陈月巧 摄)

Picture 5-9-3 Miao's Woman Festival Costume at Zhuchang of Zhijin County (Photographed by Chen Yueqiao)

第六章
六盘水地区少数民族服饰

Chapter 6 Ethnic Group's Costumes and Ornaments in Liupanshui Area

» 名称：六枝二塘苗族翻领式贯首女套装
» 时代：约1970年
» 馆藏：W.1767

主要流行于贵州六枝二塘乡，以二塘、鱼塘等地为代表。该套装由上衣、百褶裙、围腰、绑腿、头帕等组成。上衣为蜡染、挑花翻领式贯首衣，前片齐胸，后片齐腰，前胸挑四块八角花纹，后背蜡染象征蚩尤九黎的印章，后腰部蜡染、挑花图案象征祖先居住的城池和整齐的街道，以此怀念涿鹿之战中失败的蚩尤九黎和祖先被毁的家园，双袖挑花八角花纹。下装为高腰蜡染百褶长裙，裙上手绘多条蜡染横条，记录了迁徙途中渡过的许多条江河。另应有白色家织布镶边的黑色素围腰和青布绑腿。该支系也称为“四印苗”，即前胸后背及两肩袖上的四个方块即是蚩尤印，也是古代蚩尤九黎古国权力的象征。盛装时，女子佩戴银项圈、银耳环、银手镯等。

图6-1-1　六枝二塘苗族翻领式贯首女套装（正面）
Picture 6-1-1　Miao's Woman Pullover with Turn-down Collar at Ertang of Liuzhi (Front)

» Name: Miao's Woman Pullover with Turn-down Collar at Ertang of Liuzhi

» Time: About 1970

» Collection: W.1767

Represented by the style at towns of Ertang and Yutang, the outfit is popular in Ertang Town of Liuzhi Special District in Guizhou Province. It is composed of jacket, straight skirt, apron, leggings and headscarf. The batik pullover with cross-stitched collar has a front lapel with four pieces of octagonal patterns at chest length, a back lapel with batik seal representing Chiyou (Miao's warrior) at waist length, batik and cross-stitched patterns of towns, walls and streets on back waist and cross-stitched octagonal patterns on sleeves. The batik pleated skirt at waist length has many batik horizontal bars representing the rivers ancestors passed during migration. Women wear black apron trimmed with white home-woven cloth and black-cloth leggings. The branch of Miao is also called "Siyin Miao" (four-seal Miao) with the features of four seal patterns (Chiyou seal) representing the power of Chiyou in ancient times on front and back lapels and sleeves. Women wear silver ornaments including neck-rings, earrings and bracelets.

图6-1-2 六枝二塘苗族翻领式贯首女套装（背面）

Picture 6-1-2 Miao's Woman Pullover with Turn-down Collar at Ertang of Liuzhi (Back)

图6-1-3 挑花八角纹

Picture 6-1-3 Cross-stitched Octagonal Patterns

图6-1-4 螺丝纹和三角纹

Picture 6-1-4 Screw Patterns and Triangle Patterns

图6-1-5　蜡染蚩尤印章

Picture 6-1-5　Batik Chiyou Seal

图6-1-6　蜡染城郭

Picture 6-1-6　Batik Towns and Walls

图6-1-7　四印苗女盛装

Picture 6-1-7　Siyin (with Four Seals on a Jacket) Miao's Woman Festival Costume

» 名称：六枝二塘大元堡苗族右衽女套装
» 时代：约1990年
» 馆藏：W.1775

主要流行于贵州六枝、关岭、晴隆等地，以六枝二塘大元堡为代表。该套装由上衣、直筒裹裙、围腰等组成。上衣为蓝色缎面右衽绣花衣，双袖、前衣下摆由红底轴线绣花块装饰，双袖另有黑、红、黄、绿色布拼接，衣襟栏杆花边饰之。下装为青布直筒裹裙，上以红底轴线绣花块、彩色布条及锁绣回纹拼接装饰。腰系红轴线绣拼接成“川”字的长围腰。

图6-2-1 六枝二塘大元堡苗族右衽女套装（正面）
Picture 6-2-1 Miao's Right-buttoned Woman Outfit at Dayuan Pu, Ertang of Liuzhi (Front)

» Name: Miao's Right-buttoned Woman Outfit at Dayuan Pu, Ertang of Liuzhi
» Time: About 1990
» Collection: W.1775

Represented by the style at Dayuan Pu, Ertang Town of Liuzhi County, the outfit is popular in Liuzhi Speical District, Guanling County and Qinglong County in Guizhou Province. It is composed of jacket, straight skirt and apron. Sleeves sewed with black, red, yellow and green cloths and lapels adorned with railing patterns, the blue brocaded right-buttoned jacket is axis embroidered with flower patterns with red as base on sleeves and the bottom of front of the jacket. The black-cloth straight skirt is decorated with axis embroidered flower patterns in red base, colored cloth strips and lock stitched "回" (hui) patterns. Women wear long aprons with "川" (chuan) like shape formed by red axis embroidered patterns.

图6-2-2 六枝二塘大元堡苗族右衽女套装（背面）

Picture 6-2-2 Miao's Right-buttoned Woman Outfit at Dayuan Pu, Ertang of Liuzhi (Back)

» 名称：水城南开苗族背褡式对襟女套装

» 时代：约2000年

» 馆藏：W.1779

主要流行于贵州水城、纳雍、赫章等地，以水城南开乡为代表。该套装由上衣、百褶裙、羊毛绑腿等组成。上衣为家织粗布背褡式对襟衣，背褡上以黄色为基调，布拼、挑花屋架花、荞麦花等。该服装的坎肩图案，方形象征城市和田土，红色表示田中游鱼，花纹代表田螺、星辰和树林。下装为蜡染百褶裙，裙上饰有红（未婚姑娘则无红色横杠）、深蓝两条横条，分别代表祖先迁徙过程中渡过的黄河和长江，此款裙子是妇女已婚的标志，寓意着为人母的妇女必须承担起记录和传承家族历史的责任。盛装时，未婚女子头缠高大的红线，耳饰海贝串。

图6-3-1 水城南开苗族背褡式对襟女套装（正面）

Picture 6-3-1 Miao's Waist-coat Style Symmetric Woman Outfit at Nankai of Shuicheng (Front)

» Name: Miao's Waist-coat Style Symmetric Woman Outfit at Nankai of Shuicheng

» Time: About 2000

» Collection: W.1779

Represented by the style at Nankai Town of Shuicheng City, the outfit is popular in Shuicheng City, Nayong County and Hezhang County in Guizhou Province. It is composed of jacket, pleated skirt and wool leggings. Made of home-woven cloth, the waist-coat symmetric jacket is patchwork stitched and cross-stitched roof patterns of truss flower and buckwheat flower in yellow base on the waistcoat. The square patterns on sleeveless symbolize towns and fields, red represents fish in the field, and flower patterns stand for snail, stars and woods. The batik pleated skirt is decorated with red horizontal bar (no red bar for unmarried women) and dark blue horizontal bar which respectively represent the Yellow River and The Yangtze ancestors crossed during migration. This skirt is worn by married women and implies that women as mothers must bear the responsibility of recording and inheriting family history. Unmarried women have braids into topknots with red woolen strings and wear shellfish string earrings.

图6-3-2 水城南开苗族背褡式对襟女套装（背面）

Picture 6-3-2 Miao's Waist-coat Style Symmetric Woman Outfit at Nankai of Shuicheng (Back)

图6-3-3 几何纹屋架花的背褡

Picture 6-3-3 Waistcoat with Geometric Patterns of Roof Truss Flower

图6-3-4　局部几何纹屋架花的背褡

Picture 6-3-4　Waistcoat with Geometric Patterns of Roof Truss Flower (Section)

图6-3-5　百褶裙青布局部图

Picture 6-3-5　Black Cloth of Pleated Skirt (Section)

图6-3-6　百褶裙蜡染局部图

Picture 6-3-6　Batik of Pleated Skirt (Section)

图6-3-7　水城南开女盛装

Picture 6-3-7　Miao's Woman Festival Costume at Nankai of Shuicheng

第七章 铜仁地区少数民族服饰

Chapter 7 Ethnic Group's Costumes and Ornaments in Tongren Area

» 名称：印江苗族插针绣无领大襟女套装
» 时代：约1950年
» 馆藏：W.1774

主要流行于贵州印江境内苗族聚居区，以印江为代表。该套装由上衣、阔腿裤、围腰、头帕等组成。上衣为褐色缎面大襟夹衣，双袖、门襟插针绣花鸟图案，周边以栏杆花边饰之。下装为缎面阔腿裤，裤脚口插针绣蝶恋花图案，栏杆花边饰之。缎面围腰左、右下角，分别对称插针绣盆景花鸟图，围腰上端插针绣喜鹊闹春图，寓意吉祥富贵。另配有黑色头帕和绣花鞋等。贵州印江土家族苗族自治县位于贵州省东北部，由于较早受汉文化的影响，在服饰形制上更接近于中原地区传统的大襟衣。

图7-1-1　印江苗族插针绣无领大襟女套装（正面）

Picture 7-1-1　Miao’s Collarless Woman Outfit with Big Lapel at Yinjiang (Front)

» Name: Miao's Collarless Woman Outfit with Big Lapel at Yinjiang
» Time: About 1950
» Collection: W.1774

Represented by the style at Yinjiang County, the outfit is popular in Miao-populated areas of Yinjiang County in Guizhou Province. It is composed of jacket, loose pants, apron and headscarf. Made of brown brocade, the jacket with big lapel is thrusting needle stitched with flower and bird patterns surrounded by railing patterns on sleeves and lapel. The brocaded loose pants are thrusting needle stitched with butterfly and flower patterns surrounded by railing patterns on pants hem. There are symmetric thrusting needle stitched bonsai flower and bird patterns at left and right lower corner of brocaded apron. The pattern "Magpies playing in spring" in thrusting needle stitch means auspiciousness and wealth. Women wear black headscarf and embroidered shoes. Influenced by Han culture in ancient times, the costume is tailored closely to the traditional jacket with big lapel in central plains.

图7-1-2　印江苗族插针绣无领大襟女套装（背面）

Picture 7-1-2　Miao's Collarless Woman Outfit with Big Lapel at Yinjiang (Back)

图7-1-3 插针绣蝶恋花

Picture 7-1-3 Butterfly and Flower Patterns in Thrusting Needle Stitch

图7-1-4 喜鹊闹春图

Picture 7-1-4 "Magpies Playing in Spring" Pattern

图7-1-5 盆景花鸟图

Picture 7-1-5 Pattern of Bonsai Flowers and Birds

图7-1-6 插针绣花鸟纹

Picture 7-1-6 Patterns of Bird and Flower in Thrusting Needle Stitch

图7-1-7　印江苗族女盛装

Picture 7-1-7　Miao's Woman Festival Costume at Yinjiang

第八章
黔西南地区少数民族服饰

Chapter 8 Ethnic Group's Costumes and Ornaments in South-west of Guizhou Province

» 名称：贞丰坝桥苗族数纱绣右衽女套装
» 时代：约1970年
» 馆藏：W.1762

主要流行于贵州贞丰、兴仁、关岭等地区，以贞丰坝桥为代表。该套装由上衣、百褶裙、围腰、头帕等组成。上衣为家织青布右衽衣，双袖、门襟、后背以蓝、紫色调为主，数纱绣几何纹居多。下装为青布百褶长裙，裙边以织锦带饰之。腰上系长方形数纱绣菱形方格纹长围腰，另佩戴青布带穗三角形头帕。该地区数纱绣是贵州境内苗族刺绣技艺最精细的代表之一。制作过程，妇女们通过数家织布的经纬线布局，故称为数纱绣，其工艺之精湛令人感叹。盛装时，女子佩戴六、七根刻花银排圈，以及银耳环、银手镯，头饰银花等。

图8-1-1 贞丰坝桥苗族数纱绣右衽女套装（正面）

Picture 8-1-1 Miao’s Right-buttoned Woman Outfit at Baqiao of Zhenfeng (Front)

图8-1-2　贞丰坝桥苗族数纱绣右衽女套装（背面）

Picture 8-1-2　Miao's Right-buttoned Woman Outfit at Baqiao of Zhenfeng (Back)

» Name: Miao's Right-buttoned Woman Outfit at Baqiao of Zhenfeng

» Time: About 1970

» Collection: W.1762

Represented by the style at Baqiao Town, the outfit is popular in counties of Zhenfeng, Xingren and Guanling in Guizhou Province. It is composed of jacket, pleated skirt, apron and headscarf. Made of home-woven black cloth, the right-buttoned jacket is yarn-counting stitched with geometric patterns on sleeves, lapel and back mainly in blue and purple. The black-cloth pleated skirt is decorated with brocaded hem. Women wear long rectangular apron yarn-counting stitched with diamond checkerboard patterns and black-cloth triangular headscarf with tassels. Yarn-counting stitch in this area is one of the most exquisite embroidery techniques among Miao in Guizhou Province. During the embroidery, women embroider various patterns by counting the sidelong and vertical yarns of home-made cloth, so it is called yarn-counting stitch. Women wear silver ornaments including six to seven torque necklaces, earrings, bracelets and hairpins for festival occasions.

图8-1-3　贞丰苗族女老年装（雷洪斌　摄）

Picture 8-1-3　Miao's Woman Festival Costume at Baqiao of Zhenfeng (Photographed by Lei Hongbin)

第九章
遵义地区少数民族服饰

Chapter 9 Ethnic Group's Costumes and Ornaments in Zunyi Area

» 名称：遵义习水苗族挑花对襟女套装
» 时代：约1980年
» 馆藏：W.D1

遵义地区苗族在遵义市周边以及仁怀市、习水县等地均有分布。该套服饰流行于遵义习水县，为红苗盛装。上衣为对襟衣，在门襟和双袖处以挑花几何纹为主，下装为蜡染、挑花带及红布带拼接的百褶裙，腰间系挑花几何纹小长方形围腰。盛装时，头饰盘帕坠流苏和少量银饰。

图9-1-1　遵义习水苗族挑花对襟女套装（正面）
Picture 9-1-1　Miao's Cross-stitched Symmetric Woman Outfit at Xishui of Zunyi (Front)

» Name: Miao's Cross-stitched Symmetric Woman Outfit at Xishui of Zunyi

» Time: About 1980

» Collection: W. D1

Miao people in Zunyi City are distributed in the neighboring areas of Zunyi, Renhuai and Xishui County in Guizhou Province. This collection is from Xishui County of Zunyi City and is the festival costume for Red Miao. The symmetric jacket is cross-stitched with geometric patterns on lapel and sleeves. Matched with small rectangular apron with cross-stitched geometric patterns at the waist, the batik pleated skirt is sewed with cross-stitched cloth straps and red straps. Women wear turban ornamented with tassels and less silver ornaments at festivals.

图9-1-2 遵义习水苗族挑花对襟女套装（背面）

Picture 9-1-2 Miao's Cross-stitched Symmetric Woman Outfit at Xishui of Zunyi (Back)

图9-1-3　挑花几何纹门襟

Picture 9-1-3　Lapel with Cross-stitched Geometric Patterns

图9-1-4　挑花几何纹围腰

Picture 9-1-4　Apron with Cross-stitched Geometric Patterns

致谢

《盛装里的多彩贵州》主要依托五彩黔艺博物馆馆藏编撰而成。在博物馆搬迁升级之际，出版这本小书浓缩着对旧馆的一种不舍情结，也是馆藏藏品的另一种存在。在定稿的过程中，翻阅着每张图片、每段文字，回味办馆的充实和快乐，会不自主地唤起我内心深处珍贵的记忆。一种强烈的情感油然而生，那就是“感谢”。对所有帮助过我的人进行“致谢”，是件多么重要而美好的事情。

我不能忘记，当地政府和各级领导给予的全方位关心和帮助。正是因为你们在博物馆策划、筹备、建设和运行中的大力支持，才使五彩黔艺这座公益性博物馆由无到有，从小到大，茁壮成长并迸发出生机活力，成为展示贵州苗绣文化的一个重要窗口和多彩名片。

我不能忘记，尊师挚友给予的亲力投入和宝贵指导。王皞、方枚、郭红杨等老师专程从北京赶来，现场对博物馆装修风格给予艺术性的指导。贵州财经大学外语学院的张春副教授，揽下了博物馆的所有英文翻译工作，让外宾能对馆藏有更精准的认知。特别是我的博士生导师——清华大学美术学院贾京生教授全程严谨指教，让博物馆站在了一个高水准的专业平台上。

我不能忘记，博物馆团队给予的稳定的团队力量和敬业精神。张慧萍馆长勤勤恳恳，带领团队保持着博物馆的良好运行。谢庆、向宇、漆芝妤、周超、陈志航、陈月朗、朱俊敏、陈园园、徐枫等兄弟姐妹任劳任怨，像守护自己的家一样呵护着博物馆的点点滴滴。正是由于他们的精诚合作、齐心协力，五彩黔艺作为公益馆发展到今天才有了基本保障。

我不能忘记，家人和亲人给予的无私关爱和亲情付出。感谢我

的爸妈、我的先生、姐姐和姐夫、亲爱的女儿，还有众多亲朋，他们的爱，让我永远焕发出激情，永远有使不完的劲儿。他们“没有底线”的包容和支持，是我办好博物馆的动力之源。

最后，我要特别感谢贵州师范大学旅游管理学院原院长杨绍先教授。14 年前，作为我研究生的校外导师，正是他匡正了我的“寻绣”之路。清晰记得杨教授带我参观师大旅管学院的一个少数民族手工作坊，把吴晓秋老师引荐给我。正是在那里，我见识了苗绣的博大精深、妙不可言，更坚定了我的苗绣人生。

中国知名民间文艺家余未人先生拔冗作序，让本书添色增彩，也让我真切感受到余老在保护和传承民族民间文化方面的使命和责任。遵义师范学院美术学院牟孝梅教授、凯里学院的吴平教授、民间收藏家雷洪斌先生为此书提供了部分照片和有关资料，使得此书完整囊括了贵州九个地州市的服饰。有幸得益于贵阳市文旅局和观山湖区文旅局的公益资助，使这本小书顺利呈现在读者面前。

如果从收藏第一张绣片算起，到现在已经 20 年；如果从筹备五彩黔艺博物馆开始，到现在也已经 6 年。与苗绣作友为伴，是让我充实的一段五彩时光，也是让我倾情的一段美好历程。回顾来时路，整装再出发。人生事业道路上，无论走到哪里，走得多远，都不能忘记支持我走到今天的所有的人。衷心谢谢您们！

陈月巧

2021 年 8 月于青岛

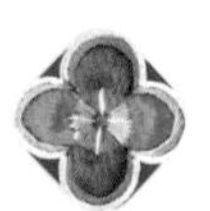

Acknowledgements

Colorful Guizhou in Festival Costumes is mainly compiled based on the collection of Guizhou Multi-color Art Museum. At the time of the relocation and upgrading of the museum, the book's publishment, another existence of the collection, condenses a kind of attachment to the old museum. In the process of finalizing the draft, looking through every picture and each paragraph of text, reminiscing about the fullness and happiness of running the museum, will involuntarily evoke the precious memories deep in my heart. A strong emotion of gratitude arises spontaneously. What an important and beautiful thing to express my thanks to all those who have helped me!

First and foremost, I would like to acknowledge the all-round care and help from the local government and leaders at all levels. It is because of their strong support in the planning, preparation, construction and operation of the museum that Guizhou Multi-color Art Museum, a public welfare museum, has grown from nothing to existence, from small to large, and burst into vitality. It has become an important window and colorful business card to display Miao embroidery culture of Guizhou.

This museum also owes a great debt to the respected teachers and best friends who have shared their hands-on input and valuable guidance. Here I would particularly like to thank teachers such as Wang Hao, Fang Mei and Guo Hongyang for coming from Beijing on a special trip to give artistic guidance on the decoration style of the museum; Zhang Chun, associate professor from School of Foreign Languages of Guizhou University of Finance and Economics, for taking over all the English translation work of the museum, so that foreign visitors can have a more accurate understanding of the collection; in particular, my

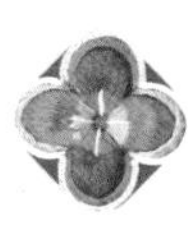

doctoral supervisor, Professor Jia Jingsheng from the Academy of Fine Arts of Tsinghua University, for giving rigorous guidance throughout the whole process to allow the museum to stand on a high-level professional platform.

I have been fortunate to have a stable team strength and professional dedication given by the museum team. The curator Zhang Huiping is very diligent, leading the team to keep the museum running well. The team members such as Xie Qing, Xiang Yu, Qi Zhiyu, Zhou Chao, Chen Zhihang, Chen Yuelang, Zhu Junmin, Chen Yuanyuan and Xu Feng, work hard and take care of every bit of the museum like guarding their own home. It is precisely because of their sincere cooperation and concerted efforts that this museum as a public welfare museum has the basic guarantee for its development.

I am grateful to selfless care and affection from my family members and relatives, especially my father and mother, my husband, elder sister and sister-in-law, my dear daughter, and many others. Their love makes me always radiate with passion and always have endless energy. Their "beyond measure" understanding and support is the source of my motivation to run the museum well.

Finally, my greatest gratitude is owed to Professor Yang Shaoxian, former dean of School of Tourism Management, Guizhou Normal University. 14 years ago, as an off-campus tutor for my postgraduate study, he was the one who rectified my "embroidery searching" path. I clearly remember that Professor Yang took me to visit a handicraft workshop for ethnic minorities in the School of Tourism Management of Normal University, and introduced Teacher Wu Xiaoqiu to me. It was there that I saw the breadth, depth and beauty of Miao embroidery, which strengthened my life in exploring Miao Embroidery.

I was pleased to invite Mr. Yu Weiren, a well-known Chinese folk artist, made a preface, adding color to the book, and also let me really feel the mission and responsibility of Mr. Yu in protecting and inheriting the national folk culture. Professor Mu Xiaomei from the School of Fine Arts of Zunyi Normal College, Professor Wu Ping from Kaili College, and private collector Lei Hongbin provided some photos and related materials

for the book, making it fully cover the costumes from nine prefectures and cities in Guizhou Province. Fortunately, thanks to the public welfare funding from Guiyang Municipal Bureau of Culture and Tourism and Bureau of Culture and Tourism of Guanshanhu District, this humble book was successfully presented to readers.

It has been 20 years since the collection of the first embroidery piece and 6 years since the preparation of this Museum. Being accompanied by Miao embroidery is not only a colorful time for me, but also a beautiful process for me. Looking back, I will start a new journey, but no matter where I go or how far I go, I will never forget all the people who have supported me till today. My sincere gratitude is for them.

Chen Yueqiao

In Qingdao, August 2021